541.3
18

£19.50

# The Tunnel Effect in Chemistry

# The Tunnel Effect in Chemistry

R.P. BELL

Hon. Research Professor of Physical Chemistry
University of Leeds

CHAPMAN AND HALL
LONDON AND NEW YORK

*First published* 1980
*by Chapman and Hall Ltd*
*11 New Fetter Lane, London EC4P 4EE*

*Published in the USA by*
*Chapman and Hall*
*an associate company of Methuen, Inc.*
*733 Third Avenue, New York, NY 10017*

© 1980 R. P. Bell

*Printed in Great Britain at the*
*University Press, Cambridge*

*ISBN 0 412 21340 0*

*All rights reserved. No part of this book may be reprinted, or reproduced or utilized in any form or by any electronic, mechanical or other means, now known or hereafter invented, including photocopying and recording, or in any information storage or retrieval system, without permission in writing from the publisher.*

**British Library Cataloguing in Publication Data**

Bell, Ronald Percy
  The tunnel effect in chemistry.
  1. Chemical reaction, Conditions and laws of
  2. Tunnelling (Physics)
  I. Title
  541'.39    QD501    79-41156

  ISBN  0 412 21340 0

# Preface

The suggestion that quantum-mechanical tunnelling might be a significant factor in some chemical reactions was first made fifty years ago by Hund, very soon after the principles of wave mechanics had been established by de Broglie, Schrödinger and Heisenberg, and similar ideas were put forward during the following thirty years by a number of authors. It was realised from the beginning that such effects would be particularly prominent in reactions involving the movement of protons or hydrogen atoms, and both theoretical and experimental work received a powerful stimulus in the discovery of deuterium in 1932. During the last twenty years theoretical predictions about the tunnel effect have been supported by an increasing body of experimental evidence, derived especially from studies of hydrogen isotope effects.

The present book presents an attempt to summarize this evidence and to indicate the main lines of the basic theory. Details of mathematical manipulation are restricted mainly to Chapter 2 and the Appendices, and many readers may prefer to confine themselves to the results obtained. The main emphasis has been on the kinetics of chemical reactions involving the transfer of protons, hydrogen atoms or hydride ions, although Chapter 6 gives an account of the role of the tunnel effect in molecular spectra, and Chapter 7 makes some mention of tunnelling in solid state phenomena, biological processes and the electrolytic discharge of hydrogen. Only passing references have been made to tunnelling by electrons. This is an essential feature of processes of oxidation and reduction in solution and at electrodes, and has recently come into prominence in connection with junctions between solid conductors, semiconductors and superconductors. Electron tunnelling has therefore been extensively reviewed in books and articles on solid state physics. For similar reasons the application of tunnelling theory to the emission of alpha particles from radioactive nuclei receives only a brief mention.

The approach adopted in this book is to regard the tunnel effect as a quantum correction (albeit sometimes a large one) which can be applied to conventional treatments of kinetic problems, in particular to transition state theory. This is in some senses an illogical procedure, since an *ab initio* quantum-mechanical treatment of such problems would in principle lead directly to transition probabilities without invoking tunnel corrections as a separate issue. The last few years have seen many attempts to develop fundamental theories of this kind, but so far they have been applied to only a very few reactions, notably those involving three hydrogen atoms. It is therefore difficult at present to assess their general usefulness in reaction kinetics, particularly for solution reactions. Moreover, both the basic concepts and the mathematical techniques employed are strange to the great majority of chemists, and it seems likely that the more empirical approach described here will retain its usefulness for some time to come, especially since for most reactions there is little prospect of obtaining quantitative information about the energy surfaces involved. Chapter 7 discusses some of the approximations and logical inconsistencies of the present approach, which would certainly have to be taken into account if and when more quantitative information becomes available. The book ends with a critical appraisal of the rather different approach to the theory of proton transfer processes adopted by Dogonadze and other physical chemists in the USSR.

The writing of this book was started while I was Guest Professor at the Physical Chemistry Institute of the Danish Technical University and finished during the tenure of a Research Professorship in the Physical Chemistry Department of the University of Leeds. Thanks are also due to the Leverhulme Trust for the award of an Emeritus Fellowship, which has greatly assisted in the preparation of the book, and to my wife for compiling the index and for much other help.

*Leeds, December* 1978                                                   R. P. Bell

# Contents

| | | |
|---|---|---|
| | **Preface** | *page* v |
| **1.** | **Physical principles and early history** | **1** |
| | 1.1 The nature of the tunnel effect | 1 |
| | 1.2 The de Broglie wavelength and the Heisenberg uncertainty principle | 3 |
| | 1.3 Optical analogies | 5 |
| | 1.4 Early applications of the tunnel effect | 8 |
| **2.** | **The calculation of permeabilities for one-dimensional barriers** | **12** |
| | 2.1 Models and methods of approach | 12 |
| | 2.2 Rectangular barriers | 17 |
| | 2.3 The parabolic barrier | 21 |
| | 2.4 Miscellaneous types of barrier | 27 |
| |    2.4.1 *The Eckart barrier* | 27 |
| |    2.4.2 *Triangular or wedge-shaped barriers* | 29 |
| |    2.4.3 *The double anharmonic barrier* | 32 |
| |    2.4.4 *The inverted Morse potential* | 32 |
| |    2.4.5 *The barrier* $V = V_0\{1 - |(x/b)^n|\}$ | 32 |
| |    2.4.6 *The numerical calculation of barrier permeabilities* | 34 |
| | 2.5 The BWK or semi-classical approximation | 34 |
| | 2.6 Tunnelling by bound particles | 42 |
| |    2.6.1 *Systems with a single minimum* | 42 |
| |    2.6.2 *Symmetrical double-minimum potentials* | 43 |
| |    2.6.3 *Unsymmetrical double-minimum potentials* | 49 |
| **3.** | **The application of tunnel corrections in chemical kinetics** | **51** |
| | 3.1 Barrier permeabilities for a Boltzmann energy distribution | 51 |
| | 3.2 The principles of applying a tunnel correction | 54 |
| | 3.3 The tunnel correction for a parabolic barrier | 60 |
| | 3.4 Tunnel corrections for other types of barrier | 63 |
| | 3.5 Activation energies and pre-exponential factors | 63 |

3.6 General criteria for tunnelling — 67
     3.6.1 *Tunnelling criteria for parabolic barriers* — 68
     3.6.2 *Model systems and the effect of transition state symmetry* — 70
     3.6.3 *Tunnelling from discrete energy levels* — 74
  3.7 Summary of experimental criteria for tunnelling — 75

**4. The theory of kinetic isotope effects** — 77

  4.1 The principles underlying kinetic isotope effects. Zero-point energies — 77
  4.2 The semi-classical transition state theory of isotope effects — 84
     4.2.1 *Basic expressions* — 84
     4.2.2 *Simplifications and model calculations* — 88
     4.2.3 *Comparison of all three hydrogen isotopes* — 91
     4.2.4 *The effect of transition state symmetry* — 92
     4.2.5 *Conclusions of the semi-classical theory* — 97
  4.3 Tunnel corrections to kinetic isotope effects — 98
     4.3.1 *The effect of tunnelling on isotope effects for Arrhenius parameters* — 99
     4.3.2 *Tunnel corrections for all three hydrogen isotopes* — 101
     4.3.3 *Tunnelling and the dependence of isotope effects on transition state symmetry* — 102
     4.3.4 *Isotope effects for tunnelling from discrete energy levels* — 104
     4.3.5 *Experimental criteria for tunnelling in kinetic isotope effects* — 105

**5. Experimental evidence for tunnelling in chemical reactions** — 106

  5.1 Introduction — 106
  5.2 Hydrogen atom transfers at low temperatures — 107
  5.3 Hydrogen atom transfers in simple gas reactions — 114
     5.3.1 *The reaction $H + H_2$ and its isotopic variants* — 117
     5.3.2 *Other gas reactions involving hydrogen abstraction* — 124
  5.4 A general survey of hydrogen isotope effects in solution — 127
  5.5 Comments on Tables 5.3 and 5.4 — 135
  5.6 Tunnelling by entities other than hydrogen — 140

**6. Tunnelling in molecular spectra** — 145

  6.1 Introduction — 145
  6.2 The predissociation of diatomic hydrides — 146
  6.3 The inversion of ammonia and related processes — 150
  6.4 Hindered rotations — 154
  6.5 Magnetic resonance spectra in labile systems — 156

*Contents* ix

**7. A review of the present position** — **163**

    7.1 Other manifestations of tunnelling — 163
        7.1.1 *Electrode processes* — 163
        7.1.2 *Solid state phenomena* — 164
        7.1.3 *Biological implications* — 165
    7.2 Current problems in the theory of tunnel corrections — 166
        7.2.1 *The validity of transition state theory* — 166
        7.2.2 *The applicability of one-dimensional tunnel corrections* — 169
        7.2.3 *Continuous versus discrete energy distributions* — 172
        7.2.4 *The role of the solvent in tunnelling* — 175

Appendix A. **Notes on hypergeometric functions** — **186**

Appendix B. **Permeabilities for the barrier** $V = V_0\{1 - |(x/b)^n|\}$ — **189**

Appendix C. **Derivation of the tunnel correction for a parabolic barrier** — **193**

**References** — **196**

**Author index**

**Subject index**

# 1 Physical principles and early history

**1.1 The nature of the tunnel effect**

The meaning of the term *tunnel effect* is illustrated by the two parts of Fig. 1. Fig. 1.1a shows a particle of mass $m$ and energy $W$ approaching the left-hand side of a potential energy barrier of height $E$, and serves as a simple model of a chemical or physical process involving activation energy. Fig. 1.1b shows how $G$, the probability that the particles will appear on the right-hand side of the barrier (i.e. that reaction will take place) varies with $W$ for a given barrier. $G$ is frequently described as the *permeability* of the barrier to the particle. The broken line represents the predictions of classical mechanics, and also our everyday experience with macroscopic objects. The probability is zero as long as $W < E$, rises abruptly to unity when $W = E$, and remains unity for all higher values of $W$. This result is independent of the width or shape of the barrier, and also of the mass of the particle.

On passing from classical mechanics to quantum mechanics the picture becomes less clear-cut, as is often the case, and the broken line in Fig. 1.1b is replaced by the continuous curve, the exact form of which depends not only on the height of the barrier, but also on its width and shape and on the mass of the particle. $G(W)$ still tends to zero and to unity for $W \to 0$ and $W \to \infty$, respectively, but the abrupt change in passing from $W < E$ to $W > E$ is replaced by a smooth increase.* The most striking feature of the continuous curve is that it predicts non-zero permeabilities even when $W < E$. This would represent an impossible situation in classical mechanics, since if the particle existed between the points $x = a$ and $x = b$ in Fig. 1.1a its total energy would be less than its potential energy, its kinetic energy negative, and hence its momentum and velocity both imaginary. The use of the term 'tunnelling' is a picturesque way of indicating that in quantum theory a particle can pass from $a$ to $b$ by some underhand means denied to particles which obey classical mechanics.

It is equally at variance with the classical picture that $G(W) < 1$ when

---

* For certain rather artificial types of barrier $G(W)$ shows oscillatory behaviour for $W \geq E$ (see Chapter 2). However, this behaviour is unlikely to be important in problems of chemical interest.

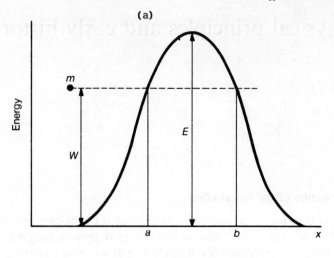

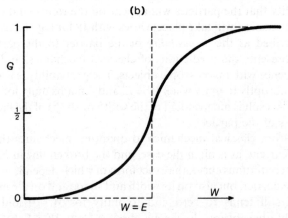

Fig. 1.1  Passage of a particle across a potential barrier in classical and quantum mechanics

$W > V$, approaching unity only at high energies. However, this behaviour is of less importance in most physical and chemical applications for two reasons. In the first place it will be shown in Section 2.4.5 and Appendix B that $G(W) \approx \frac{1}{2}$ when $W = E$ for most realistic barriers: thus $G(W)$ varies only between $\frac{1}{2}$ and 1 over the range $W = E$ to $W = \infty$, and the deviations from classical behaviour are not great. In the second place, the occurrence of states of low energy will be favoured by the Boltzmann factor $\exp(-W/kT)$, and if the temperature is not too high the existence of small but non-zero permeabilities for low values of $W$ may well make the largest contribution to the rate of the process. Nevertheless, in a quantitative treatment it may be necessary to consider the behaviour of $G(W)$ over the whole range from $W = 0$ to $W = \infty$.

## 1.2 The de Broglie wavelength and the Heisenberg uncertainty principle

The derivation from the Schrödinger equation of curves such as that in Fig. 1.1b will be deferred until Chapter 2, but their relation to the fundamental principles of the quantum theory can be explained without any detailed calculations, and it is also possible to judge under what conditions the tunnel effect will influence observable behaviour in physico-chemical systems. This is most directly seen in *the wave–particle duality of matter*. As first postulated by de Broglie [1], a particle of mass $m$ and velocity $v$ is characterized by a wavelength $\lambda$, given by the relation

$$\lambda = h/mv \tag{1.1}$$

where $h$ is Planck's constant, having the value $h = 6.626 \times 10^{-34}$ J s. As long as $\lambda$ is very small compared with the dimensions which are of interest in any particular problem the motion of the particle can be described with sufficient accuracy by classical mechanics: because of the very small value of $h$ this is always so for macroscopic systems. On the other hand, on a molecular scale both $m$ and the relevant dimensions are so small that $\lambda$ may become of the same order of magnitude as these dimensions, and when this happens considerable deviations from classical behaviour are to be expected. The nature of these deviations depends upon the type of system being considered, and in particular on whether the particle is bound or free. (For a bound particle there is a range of co-ordinates for which the potential energy is less than the total energy, so that on a classical picture the particle is confined to this region, while for a free particle this range extends to infinity in at least one direction.) For a bound system the deviations from classical behaviour take the form of quantization, i.e. the existence of discrete energy levels, while for a free particle approaching an energy barrier such deviations are manifested as the tunnel effect. Most reacting systems may, of course, involve both quantization and the tunnel effect, but it is often convenient to treat these separately.

It is therefore of interest to calculate the de Broglie wavelength $\lambda$ for particles having different masses. As an illustration we shall consider particles having a kinetic energy $E_k$ corresponding to 20 kJ mol$^{-1}$ (0.33 eV), which is comparable with the barrier heights encountered in chemical problems. Equation 1.1 can then be written in the form

$$\lambda = h(2mE_k)^{-1/2} \tag{1.2}$$

and the corresponding wavelengths for the species e$^-$, H, D, T, C and Br are given in Table 1.1.

The values of $\lambda$ in Table 1.1 may now be compared with the relevant distances in molecular systems, for example the distances traversed by atoms in a reaction, which are of the order 100 pm. It is immediately obvious that any attempt to treat the motion of electrons classically is doomed to failure, and

Table 1.1. *De Broglie wavelengths for particles of different masses*

Kinetic energy = 20 kJ mol$^{-1}$

| Particle | e$^-$ | H | D | T | C | Br |
|---|---|---|---|---|---|---|
| $m$(a.m.u.) | 1/1750 | 1 | 2 | 3 | 12 | 80 |
| $\lambda$ (pm) | 2690 | 63 | 45 | 36 | 18 | 7.0 |

this is, of course, a commonplace of theoretical chemistry. In particular, electrons should be able to tunnel through distances equal to many molecular diameters, and this is in fact the basis of many electronic devices which involve a thin dielectric layer between two solid conductors or semiconductors. At the other end of the scale it should be a good approximation to apply classical mechanics to the motion of carbon or heavier atoms, and it is rarely necessary to consider the tunnel effect when treating such processes. The isotopes of hydrogen represent an intermediate case in which $\lambda$ is of similar magnitude to molecular dimensions, and one may expect considerable deviations from classical behaviour, i.e. an appreciable tunnel effect. Since the wavelength is inversely proportional to the square root of the mass the values for H, D, and T do not differ very greatly, but we shall see later that $\lambda$ (and therefore $m^{-1/2}$) occurs exponentially in expressions for the permeability, so that the tunnel effect decreases rapidly in the series H > D > T. This is an important result, since the study of hydrogen isotope effects provides a valuable method of investigating tunnelling phenomena.

An alternative but equivalent way of relating the tunnel effect to the fundamental principles of quantum theory is through the *Heisenberg uncertainty principle* [2]. This states that it is impossible to specify simultaneously and with arbitrary accuracy the position of a particle along a given axis and its momentum along the same direction, the product of the uncertainties in these two quantities being approximately equal to $\hbar = h/2\pi$, where $h$ is Planck's constant. For motion along the $x$-axis this gives the relation

$$|\Delta x||\Delta p_x| = |\Delta x||m\Delta v_x| \approx \hbar. \tag{1.3}$$

This is clearly related to the wave picture, in which the particle appears as a wave packet, since such a representation does not correspond to a definite location and momentum, but to a spread of values for each quantity. This correspondence can be illustrated more quantitatively by considering a particle moving in one dimension in a region of constant potential energy $V_0$: since the zero of energy is arbitrary it is convenient to take $V_0 = 0$. The kinetic energy is then $E_k = \frac{1}{2}mv_x^2 = p_x^2/2m$ and the momentum is $p_x = \pm(2mE_k)^{1/2}$, i.e. the uncertainty $|\Delta p_x|$ is $2(2mE_k)^{1/2}$. Equation 1.3 then gives

$$|\Delta x| \approx \hbar/2(2mE_k)^{1/2} \approx \lambda/4\pi \tag{1.4}$$

from Equation 1.2, which is again equivalent to saying that there is an appreciable probability of barrier penetration when $\lambda$ becomes comparable with the width of the barrier.*

As already mentioned, when the wave picture of matter is applied to bound molecular systems, deviations from classical mechanics are manifested by the appearance of discrete energy levels, for example quantized vibrational energy levels. A particular result is the existence of *zero-point energy*, and this also follows directly from the uncertainty principle. A simple example is a particle confined in a one-dimensional potential box of width $a$, i.e. $V(x) = \infty$ for $x < 0$ and $x > a$, $V(x) = 0$ for $0 < x < a$. It follows immediately that the kinetic energy of such a system cannot be zero, since the position is certainly between $x = 0$ and $x = a$, while the momentum would be fixed at zero if there were no kinetic energy, and the uncertainty principle would thus be violated. If the lowest energy level is $E_0$ the momentum is $\pm(2mE_0)^{1/2}$, and Equation 1.3 gives

$$|\Delta x||\Delta p_x| = 2a(2mE_0)^{1/2} \approx \hbar$$
$$E_0 \approx \hbar^2/8ma^2 \tag{1.5}$$

which may be compared with the exact result

$$E_0 = \pi^2\hbar^2/2ma^2.$$

Similar results are obtained for a harmonic oscillator or any other potential function representing a bound particle: if we are considering vibrations of a diatomic species, $m$ is replaced by the reduced mass $\mu = m_1 m_2/(m_1 + m_2)$. We thus see that the tunnel effect and the existence of zero-point energy have exactly the same logical status in quantum theory, both being direct consequences of the uncertainty principle. It is thus illogical to consider the tunnel effect as some special or additional quantum effect, or to ignore it, if we accept the existence of zero-point energy.† One might surmise that the two types of quantum correction will often modify the results of a classical treatment to similar extents, and it will be shown subsequently that this is indeed the case.

## 1.3 Optical analogies

Because of the wave–particle duality of matter, expressed quantitatively by the de Broglie relation (Equation 1.1) and the Schrödinger equation, there are

---

* It is important to note that Equations 1.3 and 1.4 represent only approximate equalities, since it is only sometimes legitimate to specify an exact value for an uncertainty. It is therefore dangerous to attempt to derive exact relations from the uncertainty principle, though it is often useful for a semi-quantitative treatment.

† For this reason the term 'tunnel effect' is somewhat misleading, and 'tunnel correction' might be preferred: however, the former term has become generally accepted and will be used in this book in a qualitative sense, while 'tunnel correction' will be used quantitatively to describe the factor by which classical results must be corrected.

several analogies to the tunnel effect in the behaviour of electromagnetic radiation in regions where the refractive index is not uniform. One of the simplest examples is illustrated by Fig. 1.2, which shows a ray of light falling normally on a pair of 90°–45°–45° glass prisms separated by an air gap. According to ray optics the condition for total internal reflection is $n \sin \theta > 1$, where $n$ is the refractive index of the glass and $\theta$ the angle of incidence. Since $n = 1.5$–$1.9$ for various types of glass, total internal reflection should take place in the first prism, irrespective of the position of the second one. However, if Maxwell's equations for the electromagnetic theory are applied to the problem, it turns out that electromagnetic disturbances extend appreciably into the less dense medium (in this case air), though their amplitude decreases exponentially with distance. If the width of the gap between the two prisms is decreased sufficiently these disturbances can be 'caught' in the form of a transmitted ray passing through both prisms, the intensity of which decreases exponentially with increasing width of the air gap.

For the system illustrated in Fig. 1.2, if $\lambda$ is the wavelength of the radiation in air and $z$ the perpendicular distance between the prisms, the transmission coefficient or permeability $G$ (i.e. the ratio of the intensity of the transmitted beam to that of the beam incident on the boundary) is given by electromagnetic theory [3] as

$$G = [1 + A \sinh^2(bz)]^{-1} \tag{1.6}$$

where

$$A = (n^2 - 1)^2 / n^2 (n^2 - 2), \quad b = (2\pi/\lambda)(\tfrac{1}{2}n^2 - 1)^{1/2}.$$

For small degrees of transmission this becomes approximately

$$G \approx \exp\left[\frac{4\pi z}{\lambda}(\tfrac{1}{2}n^2 - 1)^{1/2}\right] \tag{1.7}$$

showing that transmission will be appreciable only when the gap width $z$ becomes comparable with the wavelength $\lambda$. This is entirely analogous to the

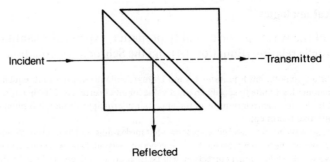

Fig. 1.2 Optical analogue of the tunnel effect

# Physical principles and early history

statement in the last section that tunnelling of particles is appreciable only when their de Broglie wavelength is comparable with the width of the relevant energy barrier, and it will be shown in Chapter 2 that the equations for transmission coefficients are essentially the same in the two cases.

This phenomenon, sometimes described as *frustrated total reflection*, was first observed by Newton in his experiments on Newton's rings, when he noticed that the area of contact between a plane and a convex surface appeared to be greater than would be expected on geometrical grounds [4]. At that time no satisfactory explanation was of course available either for the diffraction rings or for the transmission phenomena, but subsequently both types of behaviour provided conclusive quantitative evidence for the electromagnetic theory of light. The type of observation illustrated in Fig. 1.2 is well suited as a class experiment, and several up-to-date versions have been described recently [5–7]. When visible light is used the gap between the prisms has to be 1 $\mu$m or less, necessitating rather fine adjustments, but if microwaves are used transmission is appreciable with gaps of several centimetres, and 3 cm waves have been commonly used for this purpose, with paraffin wax prisms and crystal detectors [8].

A particularly simple qualitative demonstration of frustrated internal reflection, resembling Newton's original observation, has been described by Pohl [9] and is shown in Fig. 1.3.

If a knife blade is pressed against the outside of a glass of water and viewed through the water from above, much more of it is visible than is in actual contact with the glass. We can assume as a rough approximation that transmission is possible through a gap $z$ one wavelength thick (cf. Equation 1.7), say 0.6 $\mu$m. If the radius of the glass is $R = 35$ mm, simple geometry shows that the radius of the visible spot $x$ is given by

$$x \approx (2Rz)^{1/2} \approx 0.2 \text{ mm} \tag{1.8}$$

which can easily be seen.

The appearance of a transmitted beam of finite intensity is not the only phenomenon associated with frustrated internal reflection. A detailed treatment shows that both the

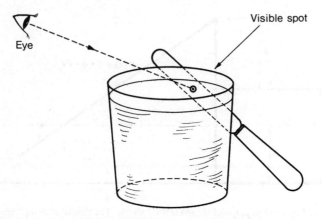

Fig. 1.3 Demonstration of the optical analogue of the tunnel effect

transmitted and the reflected beams also undergo a lateral displacement: this is the so-called Goos–Hänchen effect [10–12], which is responsible for the ease with which waves spread along the surface between two media, and is involved in such diverse phenomena as the propagation of radio waves by reflection from the ionosphere, the use of fibre optics in medicine, and the Schmidt lateral wave in seismology. The implications of this effect for tunnelling in chemical processes are not clear, although some attempts have been made to pursue the analogy for simple model systems [13].

## 1.4 Early applications of the tunnel effect

The simplest real system which illustrates the tunnel effect is the field ionization of excited states of the hydrogen atom [14], but the earliest application which permitted comparison with experiment was the treatment by Fowler and Nordheim [15, 16] of the *field-assisted emission of electrons by metals*. The situation is shown schematically in Fig. 1.4. The energy of the emitted electrons, which can be determined by measuring their velocity some distance outside the metal, is found to be much less than the value $\zeta + \phi$ which would be classically necessary to surmount the energy barrier presented by the work function $\phi$. The electron must therefore tunnel through the barrier, as indicated by the arrows in the diagram. The barrier width ($\Delta x$ in Fig. 1.4) can easily be calculated: for example if $\phi = 4$ eV and the field $F$ is $10^{10}$ V m$^{-1}$, which corresponds to a readily measurable electron emission, then $\Delta x$ must be at least $\phi/F = 4 \times 10^{-10}$ m. These large tunnelling distances are, of course, due to the low mass of the electron. Fowler and Nordheim [15, 16] showed that a quantitative treatment of the scheme shown in Fig. 1.4 gives expressions which agree with experiment, and in particular predict that the dependence of the emission $j$ upon the field strength $F$ and the electron energy $E$ should be

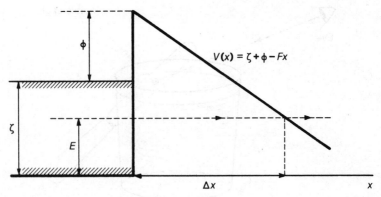

Fig. 1.4 Field-assisted emission of electrons from metals. The shaded area represents the Fermi band

## Physical principles and early history

approximately of the form

$$j = A \exp[-B(\zeta + \phi - E)^{3/2}/F] \tag{1.9}$$

where $A$ and $B$ are constants for a given metal. The derivation of transmission coefficients for this type of barrier will be described briefly in Section 2.4.2.

The theory of electron tunnelling from metals into a vacuum has subsequently been refined and extended to various types of junction between metals, dielectrics, semiconductors and superconductors. Many of these phenomena are of practical importance in electronic devices, and there is now an extensive literature on electron tunnelling in solids, including several books [17–20].

Only one application, of chemical interest, will be mentioned here. When electrons tunnel through a layer of dielectric (for example aluminium oxide) between two metal surfaces they can lose energy to vibrational modes of molecules adsorbed in the dielectric layer. A plot of electron current against applied potential will then show changes of slope at potentials for which $eV = h\nu$, where $\nu$ is a vibrational frequency, and these *inelastic electron tunnelling spectra* can thus be used to obtain information about the binding of adsorbed species. There are no selection rules governing the excitation of vibrational modes, and the method has already been applied to a number of systems [21–25].

Almost simultaneously with the interpretation of field-assisted electron emission by metals, the concept of tunnelling was applied independently by Gamow [26] and by Gurney and Condon [27, 28] to explain the *emission of α-particles by radioactive nuclei*. Although the mass of the α-particle is about 7000 times that of the electron, this is compensated for by the very small distances (about $10^{-14}$ m) which are relevant in nuclear phenomena, and tunnelling processes are again dominant. The potential energy of the α-particle as a function of its distance from the centre of the nucleus is shown schematically in Fig. 1.5. At moderate distances the energy is determined by the coulombic repulsion between the α-particle and the positive charge on the nucleus, thus decreasing hyperbolically with distance. In fact experiments on the deflection of charged particles show that the coulomb law continues to hold accurately down to very short distances; for example, for the uranium isotope $^{238}$U this is the case at least down to $3.0 \times 10^{-14}$ m, at which point the energy of repulsion amounts to $1.4 \times 10^{-12}$ J (8.8 MeV). However, at some point the repulsive forces must give way to the short-range attractive forces which hold the nucleus together, and this is indicated schematically on the diagram by a vertical line at a distance $R$. The α-particle will then occupy some energy level such as $E$ in the figure. A given type of radioactive nucleus emits α-particles of a fixed energy, and it is invariably found that this energy is much less than the energy of repulsion at the distance $R$. For example, $^{238}$U emits α-particles of energy $6.7 \times 10^{-13}$ J (4.2 MeV), while we have seen that the repulsive energy is

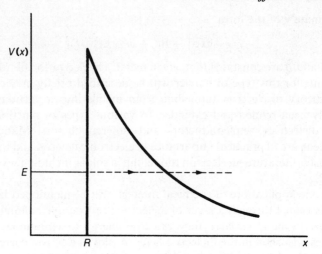

Fig. 1.5 Emission of alpha-particles from nuclei

at least $1.4 \times 10^{-12}$ J. The position thus resembles that in electron emission from metals, and it is again supposed that α-particle emission involves tunnelling through the barrier, as shown by the arrows in Fig. 1.4.

Calculation shows [26–28] that tunnelling can reasonably account for the observed rates of decay, but an exact prediction demands a knowledge of the distance $R$ and of the attractive forces inside the nucleus. A more striking achievement of the theory was its explanation of a relation first proposed empirically by Geiger and Nuttall in 1911 [29]. They showed that for a given radioactive series there is a logarithmic relation between the decay constant and the range or energy of the α-particle emitted, and a relation of this form is in fact predicted by the tunnelling theory if $R$ remains constant while the binding energy of the α-particle varies. This type of treatment has been extended in many directions. If reasonable assumptions are made about the distance $R$, tunnelling theory can be used to predict the decay constants (or half-lives) of α-emitters from the observed α-particle energies. Thus if we write

$$10^{15} R/m = 1.3 A^{1/3} + 1.20 \tag{1.10}$$

where $A$ is the atomic number, the half-lives of a wide variety of nuclei are predicted correctly to within a factor of four, although their values range over 22 powers of ten [30].

In the types of particle emission so far considered (electrons and α-particles) the process takes place from well-defined energy levels or bands and the energies and dimensions concerned are such that no classical descriptions of the phenomena are possible: we can therefore reasonably speak of 'pure tunnelling'. For *chemical reactions*, on the other hand, the position is much less clear-cut. In the first place almost all reactions involve a large number of energy levels, with a distribution depending on temperature. In the second place, Table 1.1 shows that even for the lightest atoms the effective wavelengths are not large compared with molecular dimensions, so that the tunnel effect is likely to give rise to some quantitative modifications of the classical picture rather than to qualitatively different phenomena. There were in fact until recently no observations on chemical reactions which provided compelling reasons for introducing the tunnel effect, and the detection of small quantitative effects is still hindered by lack of knowledge about the forces operating during chemical change. Nevertheless, the suggestion by Hund [31] that tunnelling might be important in chemical kinetics actually pre-dates the first treatments of particle emission [15, 26, 27], and it was followed rapidly by similar suggestions from a number of authors [32–41]. These early papers all concluded that appreciable tunnel corrections are likely only for reactions involving the movement of the ions or atoms of hydrogen. Certain predictions were made about the consequences of such corrections, particularly in causing deviations from the Arrhenius equation, and immediately after the discovery of deuterium in 1932 it was realised [39–41] that a study of hydrogen isotope effects offers the best chance of verifying such predictions, although substantial experimental evidence was not forthcoming until more than twenty years later.

The next four chapters start with a description of calculations of barrier permeabilities (Chapter 2) and their application to reaction kinetics (Chapter 3). This is followed by an account of the theory of kinetic hydrogen isotope effects, both with and without tunnel corrections (Chapter 4), and then a review of the experimental evidence for tunnelling in chemical reactions (Chapter 5).

# 2 The calculation of permeabilities for one-dimensional barriers

## 2.1 Models and methods of approach

This chapter describes the methods used for calculating permeabilities (or transmission coefficients) for a number of model situations. We shall confine our attention to the case in which the energy of the system has a single known value, leaving until Chapter 3 the problem of relating these results to the situation in chemical reactions, in which a range of energies will normally be involved. The whole treatment depends upon obtaining and interpreting solutions of the Schrödinger wave equation, for which the necessary principles and techniques are given in numerous textbooks on quantum mechanics (that by Landau and Lifshitz [42] is particularly useful in the present context). References will be given sparingly, and results often stated without a full derivation, though some of the less accessible arguments are given in more detail in the Appendices.

The model usually employed for treating tunnelling problems in chemical or physical processes consists of the motion of a particle across a potential energy barrier, and the following simplifying assumptions are commonly made:

(a) The barrier is *one-dimensional*, i.e. the potential energy $V$ is a function $V(x)$ of a single co-ordinate $x$, and the barrier can be represented by the curve obtained by plotting $V(x)$ against $x$. When a proton (or hydrogen atom) is moving between much heavier atoms $x$ will represent the position of the proton, but more generally it is natural to identify it with the *reaction co-ordinate*, which can be formulated in different ways, but is in principle a function of the positions of all the atoms which move during the reaction. In this event the one-dimensional potential energy curve should strictly speaking be replaced by a potential energy surface in several dimensions.

(b) The mass of the particle $m$ remains constant during the process (i.e. is independent of $x$). This is obviously true to the extent that the process can be regarded as the motion of a single entity, such as a proton. However, in general $m$ should be replaced by a *reduced mass* $\mu$, which is a function of the masses of several particles: the appropriate value of $\mu$ in a given process will depend upon how the reaction co-ordinate $x$ is formulated, and it will also vary with

# The calculation of permeabilities

the configuration of the system (i.e. with $x$) if large variations of the latter are involved [43].

Assumptions (a) and (b) are thus not strictly justified, but it has nevertheless been common practice to treat tunnelling problems in chemistry in terms of a one-dimensional barrier and a particle of fixed mass, the latter often being taken as the mass of a proton or deuteron. Some reference will be made in Chapters 5 and 7 to attempts which have been undertaken to treat the difficult theoretical problem of tunnelling in several dimensions, but since our knowledge of such energy surfaces is rudimentary even for the simplest chemical reactions there is some justification for retaining the simpler model.

The three types of one-dimensional barrier which occur are shown schematically in Fig. 2.1. Diagram (a) refers to what is known as *free-particle tunnelling*, since there are no minima in potential energy, corresponding to bound states, on either side of the barrier. This type of barrier might at first sight seem quite inappropriate to chemical reactions, in which bonds are broken and made. However, it must be remembered that these energy curves refer to the reacting system as a whole, and that in any bimolecular process the

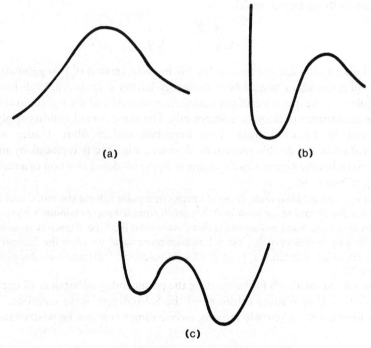

Fig. 2.1 Different types of potential energy barriers

two reacting partners are not initially bound together. For example, in a reaction of the type

$$X + YZ \rightarrow XY + Z$$

(where X, Y and Z are atoms, groups or ions) both the initial and the final states are unbound in this sense, and a treatment in terms of tunnelling by free particles may be appropriate. The relation of diagrams like Fig. 2.1a to actual chemical processes will be considered in detail in Section 3.1.

Fig. 2.1b and c show potential energy minima on one or both sides of the barrier, and therefore correspond to *tunnelling by bound particles*. Fig. 2.1b represents the transition of a particle from a bound to an unbound state (or the reverse process), and has already been illustrated in Figs. 1.3 and 1.4 for the emission of electrons from metals or of α-particles from nuclei. The same situation arises in chemical reactions if we are considering the dissociation of a species into two fragments, or the recombination of these fragments,

i.e. $XY \rightarrow X + Y$

provided that there is an energy barrier for reaction in both directions. In Fig. 2.1c both the initial and the final states are bound. This corresponds to an intramolecular process, for example the transfer of an atom Z as shown schematically in Equation 2.1.

$$C_Y^{X-Z} \rightarrow C_{Y-Z}^{X} \tag{2.1}$$

It will also be the correct picture for the interconversion of two geometrical forms of a species in cases where an energy barrier is involved. Well-known examples are the inversion of the ammonia molecule and the interconversion of the conformers of alicyclic compounds. The same considerations apply to reactions in the solid state or in immobile surface films. Under some circumstances the double minimum shown in Fig. 2.1c is replaced by more than two adjacent minima, as for example in the hindered rotation of a methyl group (Chapter 6).

In a reaction between ionic or polar species in a polar solvent the initial and final states are also bound, in the sense that they are in potential energy minima with respect to solvent motion. Some authors have therefore treated this type of reaction in terms of tunnelling by bound particles, but it has been more usual to retain the free-particle treatment, which is certainly appropriate for gas reactions. This question is discussed in Section 7.2.4.

The various methods for calculating the permeability of barriers all depend on exact or approximate solutions of the Schrödinger wave equation. The wave function for a particle moving in one dimension can be written in the form

$$\psi(x, t) = \psi(x) \exp(-iWt/\hbar) \tag{2.2}$$

# The calculation of permeabilities

where $W$ is the energy, $\hbar = h/2\pi$, and the time-independent wave function $\psi(x)$ satisfies the equation

$$\frac{d^2\psi(x)}{dx^2} + \frac{2m}{\hbar^2}[W - V(x)]\psi(x) = 0. \tag{2.3}$$

There are then two essentially different methods of applying these wave functions to the tunnelling problem. The first is the *steady state method*, which is particularly applicable to free-particle tunnelling. It pictures a constant stream of particles of energy $W$ approaching the left-hand side of the barrier, as in Fig. 2.2. Some of these will be reflected at the point $x_1$, producing a flow of particles in the opposite direction, but some will penetrate the barrier, and will appear at $x > x_2$ as a flow from left to right having a current density less than that of the stream incident at $x = x_1$. Since the whole system is in a steady state we need only consider the time-independent wave functions, the general form of which in different regions of the diagram can be inferred from Equation 2.3 without any detailed knowledge of the shape of the barrier. This is illustrated in the lower part of Fig. 2.2, which shows the behaviour of $\psi\psi^*$, which is

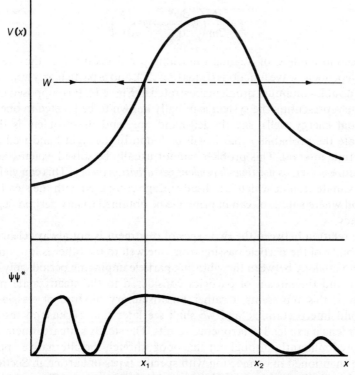

Fig. 2.2 Wave functions for a particle tunnelling through a potential barrier

proportional to the probability of finding a particle at any point. In the free-particle regions $x < x_1$ and $x > x_2$, where $W - V(x)$ is positive, $\psi\psi^*$ will oscillate with a wavelength which decreases with increasing values of $W - V(x)$, but in the classically forbidden region $x_1 < x < x_2$, where $W - V(x)$ is negative, it will not be oscillatory in character, and will approximate to an exponential increase or decrease with distance. Since the Schrödinger equation is a second-order differential equation each solution will contain two adjustable parameters, and the calculation of the permeability consists essentially in finding solutions which have the correct form for $x \ll x_1$ and $x \gg x_2$, and which remain finite and continuous throughout.

The permeability is then found by comparing the wave functions at $x < x_1$ and $x > x_2$. The value at any point of $\psi\psi^*$ (where $\psi^*$ is the complex conjugate of $\psi$) is proportional to the probability of finding a particle between $x$ and $x + dx$, and is termed the *particle density* at that point. In tunnelling problems it is often more useful to consider the *particle flux j*, i.e. the number of particles passing a given point in unit time. This can be obtained by combining the particle density with the particle velocity, and for motion in one dimension is given by the general expression

$$j = \frac{i\hbar}{2m}\left(\psi \frac{d\psi^*}{dx} - \psi^* \frac{d\psi}{dx}\right). \tag{2.5}$$

The second method of treating tunnelling problems involves the use of the time-dependent wave functions (Equation 2.2), and is particularly appropriate to the double-minimum problem illustrated in Fig. 2.1c. It is supposed that the particle representing the system is initially known to be located in one of the potential energy wells, say the left-hand one, and the problem is then to calculate the probability that it will be found in the right-hand well after a given time interval. This problem cannot usually be solved exactly, and the procedure is then to use *time-dependent perturbation theory*. This can yield only approximate results, which is a disadvantage compared with the steady state method where solutions can in principle be obtained to any desired degree of accuracy.

The relation between the two types of treatment is not always clear. If the probability of the particle passing from one well to the other is low, one may draw an analogy between the vibrating particle impinging periodically on the barrier and the stream of particles considered in the steady state picture. However, this represents a rather unsatisfactory mixture of classical and quantum interpretations, and we shall see later that because of resonance phenomena it can lead to erroneous results. The steady state treatment will be used predominantly throughout this book, though the alternative approach will be mentioned in connection with specific types of barrier, in Section 2.6, and when dealing with tunnelling in molecular spectra in Chapter 6.

## 2.2 Rectangular barriers

Although a rectangular barrier is obviously a poor approximation to most physical situations, because of the discontinuities in potential energy which it implies, we shall follow precedent in starting with a fairly detailed discussion of this type of barrier. This course has been adopted because the relevant exact wave functions can all be expressed in terms of elementary functions, which makes it easier for the average chemist to follow the steps in the argument and to appreciate the relation between different types of exact and approximate treatment.

The general barrier of this type is shown in Fig. 2.3, in which the regions of constant potential energy $V_1$ and $V_3$ extend to $x = -\infty$ and $x = +\infty$ respectively. It is then easily verified that when $W < V_2$ Equation 2.3 is satisfied by the following wave functions in the three regions

$$
\begin{aligned}
\text{I} \quad & x < 0, \quad \psi_1 = e^{ik_1 x} + B_1 e^{-ik_1 x} \\
\text{II} \quad & 0 < x < a, \quad \psi_2 = A_2 e^{\kappa_2 x} + B_2 e^{-\kappa_2 x} \\
\text{III} \quad & x > a, \quad \psi_3 = A_3 e^{ik_3 x} + B_3 e^{-ik_3 x}
\end{aligned}
\qquad (2.6)
$$

in which

$$k_1^2 = \frac{2m}{\hbar^2}(W - V_1), \quad \kappa_2^2 = \frac{2m}{\hbar^2}(V_2 - W), \quad k_3^2 = \frac{2m}{\hbar^2}(W - V_3). \qquad (2.7)$$

$k_1$, $\kappa_2$ and $k_3$ are thus all real positive quantities. Since only the ratios of the coefficients are significant the coefficient of $e^{ik_1 x}$ in $\psi_1$ has been taken as unity, while the remaining coefficients are in general complex numbers.

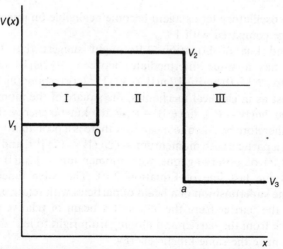

Fig. 2.3 The unsymmetrical rectangular potential barrier

The wave functions in regions I and III have a simple physical interpretation. Thus the particle density in the first region is given as

$$\psi_1 \psi_1^* = (e^{ik_1 x} + B_1 e^{-ik_1 x})(e^{-ik_1 x} + B_1^* e^{ik_1 x})$$
$$= 1 + |B_1|^2 + 2f \cos 2k_1 x + 2g \sin 2k_1 x \qquad (2.8)$$
$$= 1 + |B_1|^2 + 2|B_1| \cos(2k_1 x - \delta)$$

where $B_1$ is the complex number $f + ig$, $B_1^2 = B_1 B_1^*$, and $\tan \delta = g/f$. This is an oscillatory function, but its mean value over a distance which is large compared with $1/k_1$ is the constant quantity $1 + |B_1|^2$. In quantum theory the observed value of the momentum is obtained by replacing $p_x$ by the operator $(\hbar/i)d/dx$, i.e.

$$p_x = \int \psi^* \frac{\hbar}{i} \frac{d\psi}{dx} dx \bigg/ \int \psi \psi^* dx. \qquad (2.9)$$

This gives for region I

$$p_x = \frac{\hbar k_1 \int \{1 - |B_1|^2 + 2|B_1| \sin(2k_1 x - \delta)\} dx}{\int \{1 + |B_1|^2 + 2|B_1| \cos(2k_1 x - \delta)\} dx}$$

$$= \frac{1 - |B_1|^2}{1 + |B_1|^2} [2m(W - V_1)]^{1/2} \qquad (2.10)$$

$$= \frac{1}{1 + |B_1|^2} [2m(W - V_1)]^{1/2} - \frac{|B_1|^2}{1 + |B_1|^2} [2m(W - V_1)]^{1/2}$$

in which the oscillatory terms again become negligible on integrating over a distance large compared with $1/k_1$.

The second line of Equation 2.10 might suggest that the observed momentum has a value intermediate between $+[2m(W - V_1)]^{1/2}$ (corresponding to $|B_1| = 0$) and $-[2m(W - V_1)]^{1/2}$ (corresponding to $|B_1| = \infty$). However, just as in classical mechanics, the square of the momentum has a definite value $2m(W - V_1)$, since $W - V_1$ is the kinetic energy. The quantum result must therefore be taken to mean that there is a probability $1/(1 + |B_1|^2)$ of observing a particle with momentum $+[2m(W - V_1)]^{1/2}$, and a probability $|B_1|^2/(1 + |B_1|^2)$ of observing one with momentum $-[2m(W - V_1)]^{1/2}$, as indicated in the last line of Equation 2.10. The wave function $\psi_1$ thus represents the superposition of a beam of particles with relative density unity incident on the barrier from the left, and a beam of relative density $|B_1|^2$ reflected back from the barrier and moving from right to left: the particles in both beams have the same kinetic energy.

Similar remarks apply to the interpretation of $\psi_3$ in region III. In the present

## The calculation of permeabilities

context we are supposing that all particles which cross the barrier from left to right disappear in a positive direction, so that there are no particles moving from right to left in this region. We must therefore take $B_3 = 0$, reducing the number of unknown coefficients to four; $B_1, A_2, B_2$ and $A_3$. The values of these coefficients can be found from the four equations expressing the continuity of $\psi$ and $d\psi/dx$ at the points $x = 0$ and $x = a$, which are given in Equation 2.11.*

$$1 + B_1 = A_2 + B_2$$
$$ik_1(1 - B_1) = \kappa_2(A_2 - B_2)$$
$$A_2 e^{\kappa_2 a} + B_2 e^{-\kappa_2 a} = A_3 e^{ik_3 a} \qquad (2.11)$$
$$\kappa_2 A_2 e^{\kappa_2 a} - \kappa_2 B_2 e^{-\kappa_2 a} = ik_3 A_3 e^{ik_3 a}.$$

On eliminating $B_1$, $A_2$ and $B_2$ from these equations, we find for $A_3$

$$A_3 = 2e^{-ik_3 a} \bigg/ \left[ \left(1 + \frac{k_3}{k_1}\right) \cosh \kappa_2 a + i\left(\frac{\kappa_2}{k_1} - \frac{k_3}{\kappa_2}\right) \sinh \kappa_2 a \right]. \qquad (2.12)$$

Since the particle density in the incident beam has been taken as unity (by putting $A_1 = 1$ in Equation 2.6), the value of $A_3 A_3^* = |A_3|^2$ represents the ratio between the particle densities in the transmitted and incident beams. The permeability $G$, defined as the ratio between the particle fluxes in these two beams, is obtained by multiplying $|A_3|^2$ by the ratio of the corresponding particle velocities, which is $k_3/k_1$, giving finally

$$G = \frac{k_3}{k_1} |A_3|^2$$
$$= 4\frac{k_3}{k_1} \bigg/ \left[ \left(1 + \frac{k_3}{k_1}\right)^2 \cosh^2 \kappa_2 a + \left(\frac{\kappa_2}{k_1} - \frac{k_3}{\kappa_2}\right)^2 \sinh^2 \kappa_2 a \right]. \qquad (2.13)$$

The factor $e^{-ik_3 a}$ in Equation 2.12 disappears in $|A_3|^2$, and the same would be true in calculating any observable quantity, since the relevant expressions always contain a product of $\psi$ and $\psi^*$ or derivatives of these functions (cf. Equations 2.5 and 2.9). This factor could therefore be replaced by any quantity of modulus unity, and it is often omitted when specifying $A_3$.

Expressions can also be obtained from Equation 2.11 for the coefficients $B_1, A_2, B_2$, and in particular we find

$$B_1 = \frac{2 \cosh \kappa_2 a - (2ik_3/\kappa_2)\sinh \kappa_2 a}{(1 + k_3/k_1)\cosh \kappa_2 a + i(\kappa_2/k_1 - k_3/\kappa_2)\sinh \kappa_2 a} - 1. \qquad (2.14)$$

By comparing Equations 2.12 and 2.14 it is easily shown that

$$k_1 |B_1|^2 + k_3 |A_3|^2 = k_1. \qquad (2.15)$$

---

* Although $V(x)$ in the present problem contains finite jump discontinuities, it can be shown that this does not lead to discontinuities in either $\psi$ or $d\psi/dx$. However, if infinite jump discontinuities are present [i.e. if $V(x)$ goes suddenly to $\pm \infty$ at any point], $d\psi/dx$ is no longer continuous, though $\psi$ remains so since it must be single-valued.

Since $k_1$ and $k_3$ are proportional to the particle velocities in regions I and III respectively, Equation 2.15 means that the particle flux in the incident beam is equal to the sum of the fluxes in the reflected and transmitted beams, as is physically necessary.

For the region $0 < x < a$ Equations 2.11 give further

$$A_2 = \tfrac{1}{2} A_3 e^{ik_3 a}[1 + (ik_3/\kappa_2)] e^{-\kappa_2 a}$$
$$B_2 = \tfrac{1}{2} A_3 e^{ik_3 a}[1 - (ik_3/\kappa_2)] e^{\kappa_2 a} \qquad (2.16)$$

with $A_3$ given by Equation 2.12. The relative particle density in this region is then

$$\psi_2 \psi_2^* = (A_2 e^{\kappa_2 x} + B_2 e^{-\kappa_2 x})(A_2^* e^{\kappa_2 x} + B_2^* e^{-\kappa_2 x})$$
$$= |A_3|^2 \{[1 + k_3^2/\kappa_2^2] \cosh^2[\kappa_2(a-x)] - k_3^2/\kappa_2^2\}. \qquad (2.17)$$

Equation 2.17 represents an approximately exponential decrease of $\psi\psi^*$ between $x = 0$ and $x = a$, as sketched in Fig. 2.2. On the other hand, if Equation 2.9 is applied to calculate the observed momentum in this region, complex rather than real values are obtained, in agreement with the fact that $W < V$ and motion of the particle in this region is classically forbidden.

A similar procedure can be used to determine the behaviour of a stream of particles with $W > V_2$, the only difference being that the wave function in region II now contains the terms $\exp(\pm ik_2 x)$, where $k_2^2 = 2m(W - V_2)/\hbar^2$ (cf. Equation 2.7). The resulting expression for $G$ is

$$G = 4\frac{k_3}{k_1} \bigg/ \left[ \left(1 + \frac{k_3}{k_1}\right)^2 \cos^2 k_2 a + \left(\frac{k_2}{k_1} + \frac{k_3}{k_2}\right)^2 \sin^2 k_2 a \right]. \qquad (2.18)$$

Since $\cosh ik_2 a = \cos k_2 a$ and $\sinh ik_2 a = -i \sin k_2 a$, Equation 2.18 can be obtained from Equation 2.13 by replacing $\kappa_2$ by $ik_2$, though the logic of this procedure would need careful examination. In contrast to a classical treatment, which predicts a permeability of unity for all values of $W > V_2$, Equation 2.18 shows that $G$ is an oscillatory function of $W$: it does however approach unity asymptotically as $W \to \infty$, since the ratios $k_3/k_1$, $k_2/k_1$ and $k_3/k_2$ then all tend to unity. When $W = V_2$ both $\kappa_2$ and $k_2$ become zero, and Equations 2.13 and 2.18 give the common result

$$G = 4\frac{k_3}{k_1} \bigg/ \left[ \left(1 + \frac{k_3}{k_1}\right)^2 + k_3^2 a^2 \right] \quad (W = V_2) \qquad (2.19)$$

since both $\kappa_2^{-2} \sinh^2 \kappa_2 a$ and $k_2^{-2} \sin^2 k_2 a$ tend to the value $a^2$ as $\kappa_2$ or $k_2$ approaches zero. Classical mechanics makes no prediction about the permeability when the energy of the particle exactly matches the height of the barrier.

The above expressions can be simplified in the case of a symmetrical barrier, i.e. one in which $V_1 = V_3$. It is then convenient to take this as the zero of the energy scale, and to call the height of the barrier $V_0$. If we write

$$k^2 = 2mW/\hbar^2, \quad \kappa_0^2 = 2m(V_0 - W)/\hbar^2, \quad k_0^2 = 2m(W - V_0)/\hbar^2 \qquad (2.20)$$

# The calculation of permeabilities

the resulting expressions are:

$$W < V_0 \quad G = \frac{4k^2 \kappa_0^2}{(k^2 + \kappa_0^2)^2 \sinh^2 \kappa_0 a + 4k^2 \kappa_0^2}$$

$$= \frac{4W(V_0 - W)}{V_0^2 \sinh^2 \kappa_0 a + 4W(V_0 - W)} \quad (2.21)$$

$$W > V_0 \quad G = \frac{4k^2 k_0^2}{(k^2 - k_0^2)^2 \sin^2 k_0 a + 4k^2 k_0^2}$$

$$= \frac{4W(W - V_0)}{V_0^2 \sin^2 k_0 a + 4W(W - V_0)} \quad (2.22)$$

$$W = V_0 \quad G = (1 + \tfrac{1}{4}k^2 a^2)^{-1} = \left(1 + \frac{ma^2 V_0}{2\hbar^2}\right)^{-1}. \quad (2.23)$$

Equation 2.21 is very similar in form to Equation 1.6, which gives the transmission coefficient for the optical experiment illustrated in Fig. 1.2. The two equations can in fact be shown to be completely equivalent if it is remembered that the de Broglie wavelength of the particle outside the barrier is $h/mv = 2\pi/k$, while the critical condition for total internal reflection in ray optics is $n^2 \sin^2 \theta > 1$, which must be compared with the classical condition for the total reflection of the particle from the barrier, $V_0/W > 1$.

All the expressions given so far for the permeability of rectangular barriers to free particles are exact. For $W < V_2$ a useful approximation applies when $\kappa_2 a \gg 1$, i.e. for low permeabilities. Under these conditions both $\sinh \kappa a$ and $\cosh \kappa a$ are approximately equal to $\tfrac{1}{2}e^{\kappa a}$, and for a symmetrical barrier Equation 2.13 becomes

$$G \approx \frac{16W(V_0 - W)}{V_0^2} \exp\{-2[2m(V_0 - W)]^{1/2} a/\hbar\}. \quad (2.24)$$

Under conditions for which this approximation is valid the pre-exponential factor varies only slowly with $W$, and the equation thus illustrates the inverse exponential dependence of the permeability upon the width of the barrier, and upon the mass of the particle and the extent to which its energy falls short of the top of the barrier.

## 2.3 The parabolic barrier

This type of barrier is shown in Fig. 2.4, where the parabola is assumed to continue to infinity in both the positive and the negative directions. It is convenient to take the origin of the co-ordinate $x$ and of the energy at the top of the barrier, so that the potential energy is defined by Equation 2.25,

$$V(x) = -\tfrac{1}{2}Ax^2 \quad (2.25)$$

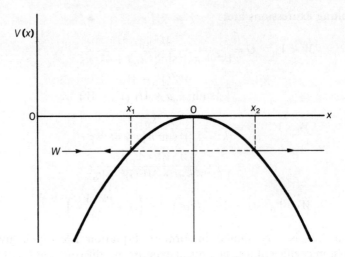

Fig. 2.4 The infinite parabolic barrier

where $A$ is a positive constant analogous to the force constant for a harmonic oscillator, for which the potential energy is represented by a parabola with the vertex downwards.

The parabolic barrier is of particular importance in chemical problems for the following reasons:

(a) It represents a realistic representation of a real barrier, at least when tunnelling involves only the upper part of the barrier, since any curve with a finite radius of curvature can be approximated by a parabola over a limited range.

(b) An exact expression can be obtained for the permeability, as shown later in this section.

(c) This expression can be integrated over a Boltzmann energy distribution to give a closed expression for the tunnel correction to the reaction velocity, as shown in Chapter 3. Most types of barrier lead to expressions for the permeability which cannot be explicitly integrated.

(d) The treatment of the reaction co-ordinate as a separable co-ordinate is strictly legitimate only over a region in which it can be represented by a parabola. This is analogous to the familiar problem of the normal vibrations of a molecule, which are separable only when they are harmonic. This problem will be discussed further in Chapters 5 and 7.

With the potential energy given by Equation 2.25, the Schrödinger equation becomes

$$\frac{d^2\psi}{dx^2} + \frac{2m}{\hbar^2}(W + \tfrac{1}{2}Ax^2)\psi = 0. \qquad (2.26)$$

By making the substitutions of Equation 2.27, Equation 2.26 is converted into

## The calculation of permeabilities

Equation 2.28,

$$\xi = x(mA/\hbar^2)^{1/4}, \quad \lambda = 2m^{1/2}W/\hbar A^{1/2} \tag{2.27}$$

$$\psi'' + (\lambda + \xi^2)\psi = 0 \tag{2.28}$$

in which the prime denotes differentiation with respect to $\xi$. The solutions of this equation have long been known under the name of Weber functions or parabolic cylinder functions, and their properties can be used to derive an expression for the permeability [44]. This involves a somewhat complicated procedure, and the same result can be obtained by a much simpler method without formulating the exact wave functions.

For sufficiently large values of $|\xi|$ the term $\lambda\psi$ in Equation 2.28 can be neglected, giving $\psi'' = -\xi^2\psi$. Under these conditions asymptotic solutions of the equation are (omitting a normalization factor)

$$\psi_\pm = e^{\pm i\xi^2/2} \tag{2.29}$$

since direct differentiation gives $\psi''_\pm = (\pm i - \xi^2)\psi$, in which the first term can be neglected in the approximation which we are using. The general asymptotic solution will then be some linear combination of $\psi_+$ and $\psi_-$.

It should be noted that when $\xi$ (and therefore $x$) is positive, $\psi_+$ and $\psi_-$ correspond to streams of particles moving in a positive and a negative direction respectively, but that when $\xi$ is negative this correspondence is reversed. This is most readily shown by applying Equation 2.5 for the particle flux, when we find $j_\pm = \pm x(A/m)^{1/2}$ for the two wave functions. The same correspondence applies to the more accurate asymptotic wave functions given in Equation 2.30. This behaviour contrasts with that found for a particle moving in a region of constant potential energy, when $e^{ikx}$ and $e^{-ikx}$ (with $k^2 = 2m(W-V)/\hbar^2$) always represent streams of particles moving in the positive and negative directions respectively, irrespective of the sign of $x$.

More accurate wave functions for large values of $|\xi|$ can be obtained by exploring solutions of the type

$$\psi_\pm = e^{\pm i\xi^2/2} f_\pm(\xi). \tag{2.30}$$

On inserting Equation 2.30 in Equation 2.28, we find that $f_\pm$ must satisfy the equations

$$f''_\pm \pm 2i\xi f'_\pm + (\lambda \pm i)f_\pm = 0. \tag{2.31}$$

For sufficiently large values of $|\xi|$ the first term of Equation 2.31 can be neglected compared with the second, and hence

$$\frac{f'_\pm}{f_\pm} = \frac{d\ln f_\pm}{d\xi} = \frac{-\tfrac{1}{2} \pm \tfrac{1}{2}i\lambda}{\xi} \tag{2.32}$$

giving on integration*

$$\ln f_\pm = (-\tfrac{1}{2} \pm \tfrac{1}{2}i\lambda)\ln \xi, \quad f_\pm = |\xi|^{-(1/2) \pm (i\lambda/2)}. \tag{2.33}$$

* No integration constant is necessary, since this would only introduce a multiplying factor in Equation 2.30.

The asymptotic wave functions for large $|\xi|$ are therefore given by Equations 2.34 and 2.35,

$$\psi = e^{-i\xi^2/2}|\xi|^{-(1/2)-(i\lambda/2)} + Ae^{i\xi^2/2}|\xi|^{-(1/2)+(i\lambda/2)} \quad (\xi \to -\infty) \quad (2.34)$$

$$\psi = Be^{i\xi^2/2}\xi^{-(1/2)+(i\lambda/2)} \quad (\xi \to +\infty). \quad (2.35)$$

The coefficient of the first term of Equation 2.34, which represents a stream of particles incident on the barrier from the left, is arbitrarily taken as unity, while Equation 2.35 contains no contribution from $\psi_-$, since we are assuming that no particles are incident on the barrier from the right.

We now need a relation between the coefficients $A$ and $B$, but this cannot be obtained by joining Equation 2.34 and 2.35 along the real axis of $x$, since they are not valid for small values of $|\xi|$. In particular, at the points marked $x_1$ and $x_2$ in Fig. 2.4 $W - V$, and hence $\lambda + \xi^2$, changes sign, and solutions of the type in Equation 2.30 are clearly invalid. The difficulty can be circumvented by regarding $\xi$ as a complex variable $u + iv$, which can be represented in the usual way in two dimensions, as shown in Fig. 2.5. The two real values $\xi = +\rho$ and $\xi = -\rho$ can be connected by the semicircular path in the upper half-plane of the diagram.

Provided that $\rho$ is sufficiently great the second term of Equation 2.34 is the dominant one throughout the path, and this term must therefore be converted into Equation 2.35 on traversing the semicircle. Since a change of sign in $\xi$ corresponds to multiplication by $e^{\pi i}$, this gives for the relation between $A$ and $B$

$$A = B(e^{\pi i})^{-(1/2)+(i\lambda/2)} = Be^{-\pi i/2}e^{-\pi\lambda/2} = -iBe^{-\pi\lambda/2}. \quad (2.36)$$

Further, since the total number of particles must be conserved,*

$$|A|^2 + |B|^2 = 1 \quad (2.37)$$

and on combining Equations 2.36 and 2.37 we obtain finally for the permeability $G$

$$G = |B|^2 = (1 + e^{-\pi\lambda})^{-1}. \quad (2.38)$$

Although the derivation of Equation 2.38 has involved approximate and asymptotic solutions, it is in fact *exact* for all values of $\lambda$ (and hence $W$), since the radius $\rho$ of the semicircle can be made as large as desired, and hence the errors involved can be made arbitrarily small. This state of affairs is peculiar to the parabolic barrier, though we shall see in Section 2.5 that a similar procedure can be used to obtain approximate expressions for the permeability of other types of barrier.

Since the origin of energy has been taken at the top of the barrier, positive

---

* Because of the symmetry of $V(x)$ the particle velocities are the same at the points $\xi = \pm\rho$, and there is no need to multiply the particle densities by the velocities in order to obtain the fluxes.

# The calculation of permeabilities

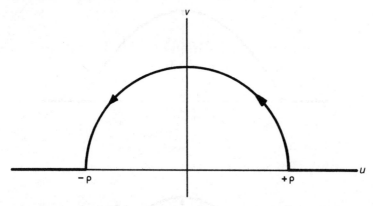

Fig. 2.5 Joining wave functions for the parabolic barrier by a contour in the complex $u$, $v$-plane

and negative values of $\lambda$ correspond respectively to particles passing 'over' and 'through' the barrier, and Equation 2.38 is equally valid in both cases. When the energy of the particle exactly matches that of the top of the barrier $\lambda = 0$ and $G = \frac{1}{2}$. If the potential energy at this point is termed $V_0$, Equation 2.38 can be written in the form

$$G = \{1 + \exp[2\pi m^{1/2}(V_0 - W)/\hbar A^{1/2}]\}^{-1} \quad (2.39)$$

by substituting for $\lambda$ from Equation 2.27.

It is often convenient to express the properties of a parabolic barrier in terms of the so-called *imaginary frequency* $iv_\ddagger$, where $v_\ddagger$ is the frequency of oscillation of a particle of mass $m$ in a parabolic potential well having the same curvature $A$ as the parabolic barrier. Since $v_\ddagger$ is given by

$$v_\ddagger = \frac{1}{2\pi}\left(\frac{A}{m}\right)^{1/2} \quad (2.40)$$

Equation 2.39 becomes

$$G = \{1 + \exp[(V_0 - W)/\hbar v_\ddagger]\}^{-1} \quad (2.41)$$

which can also be written in terms of Planck's constant as originally defined, $h = 2\pi\hbar$, or the angular frequency $\omega_\ddagger = 2\pi v_\ddagger$.

The origin of the term *imaginary frequency* is as follows. In solving the equations of motion for a bound system of particles, the frequencies are given in terms of the roots of the secular equation $\lambda_i = 4\pi^2 v_i^2 \mu_i$. For a stable molecule all these roots are positive, corresponding to real vibrations, but for a transition state there is one negative root, corresponding to motion along the reaction co-ordinate. This root, which defines the curvature of the energy barrier, corresponds formally to an imaginary frequency.

The above treatment assumes throughout that the potential energy is given by Equation 2.25 over the whole range from $x = -\infty$ to $x = +\infty$, i.e. that we

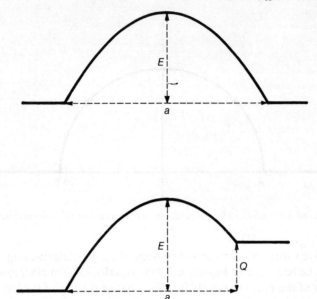

Fig. 2.6 Symmetrical and unsymmetrical truncated parabolic barriers

are dealing with an *infinite parabolic barrier*. In practice the range of possible energies is finite, and a more realistic model is provided by the *truncated parabolic barrier*, as illustrated in Fig. 2.6. We can now ascribe to the barrier a definite height $E$ and a width $a$,* as shown in the figure, and these are related to the curvature $A$ by Equations 2.42 and 2.43 for the symmetrical and unsymmetrical barriers respectively, where $Q$ is the endothermicity of the process.

$$A = 8E/a^2 \tag{2.42}$$

$$A = 2[E^{1/2} + (E-Q)^{1/2}]^2/a^2. \tag{2.43}$$

Similar expressions follow from Equation 2.40 for the characteristic frequency $v_\ddagger$ in terms of $E$, $a$ and $Q$.

The expressions derived above for the permeability of an infinite parabolic barrier are not strictly valid for a truncated one, though the error will be small unless we are dealing with energies near the base of the barrier. It has in fact been common practice to retain these expressions, but to use the truncation to define the lower limit of energies over which the permeability must be integrated. This procedure is illustrated in Chapter 3.

* The width of parabolic or other barriers has been termed $a$ by some authors and $2a$ by others. The former definition will be used throughout this book.

# The calculation of permeabilities

## 2.4 Miscellaneous types of barrier

The principles and methods of deriving permeabilities for free particles are sufficiently well illustrated by the fairly detailed account given for rectangular and parabolic barriers in the last two sections. The present section gives a much briefer account for various other types of one-dimensional barrier for which exact solutions can be obtained. In most instances only the results and some indication of the method of solution are given.

### 2.4.1 The Eckart barrier

This was the first realistic type of barrier for which the tunnelling problem was solved [45], and the results have been extensively used in chemical applications. The general equation for the Eckart barrier is

$$V(x) = \frac{Ay}{(1+y)^2} + \frac{By}{(1+y)} \quad y = e^{x/b} \tag{2.45}$$

and the barrier is illustrated in Fig. 2.7a. This expression has limiting values of $V \to 0$ as $y \to 0$ ($x \to -\infty$) and $V \to B$ as $y \to \infty$ ($x \to +\infty$), so that $B$ represents the endothermicity of the process. When $y = (A+B)/(A-B)$, $y$ has a maximum value $V_0 = (A+B)^2/4A$, which is therefore the height of the energy barrier for motion from left to right.

When $B = 0$ the barrier becomes symmetrical with its maximum at $y = 1$ ($x = 0$), $V_0 = A/4$, as shown in Fig. 2.7b, and its equation is conveniently written as

$$V(x) = V_0 \operatorname{sech}^2(x/2b) \tag{2.46}$$

where $V_0$ is the barrier height and the zero of energy is taken as the limiting value reached on either side of the barrier.

If either Equation 2.45 or Equation 2.46 is inserted in the Schrödinger equation, exact solutions can be obtained and exact values derived for the permeability. Introducing the abbreviations

$$k = (2mW)^{1/2}/\hbar, \quad \Delta = [(W-B)/W]^{1/2} \tag{2.47}$$

the following results are obtained, provided that $8mb^2A/\hbar^2 > 1$, as is always the case in chemical applications:

Unsymmetrical barrier

$$G = \frac{\sinh^2[\pi kb(1+\Delta)] - \sinh^2[\pi kb(1-\Delta)]}{\sinh^2[\pi kb(1+\Delta)] + \cosh^2[\frac{1}{2}\pi(8mb^2A/\hbar^2 - 1)^{1/2}]} \tag{2.48}$$

Symmetrical barrier

$$G = \frac{\sinh^2 2\pi bk}{\sinh^2 2\pi bk + \cosh^2[\frac{1}{2}\pi(32mb^2V_0/\hbar^2 - 1)^{1/2}]}. \tag{2.49}$$

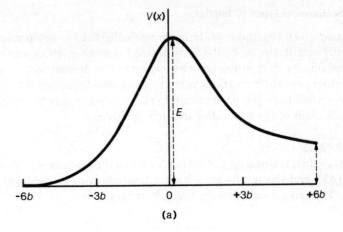

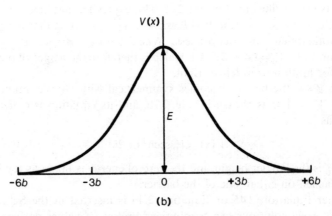

Fig. 2.7 Unsymmetrical and symmetrical Eckart barriers. The unsymmetrical curve corresponds to $Q = \tfrac{1}{3} V_0$.

These expressions are valid both for $W < V_0$ and for $W > V_0$: they tend to zero as $W \to 0$ and to unity as $W \to \infty$, as they should. They do not assume any simple form for $W = V_0$, but for barriers of dimensions characteristic for chemical reactions $G$ is then very close to one half, the exact value for an infinite parabolic barrier [cf. Equation 2.39].

The derivation of Equation 2.49 for the symmetrical barrier is in outline as follows. With the potential energy given by Equation 2.46 the Schrödinger equation becomes

$$\frac{d^2\psi}{dx^2} + \frac{2m}{\hbar^2}\left[W - V_0 \operatorname{sech}^2\left(\frac{x}{2b}\right)\right]\psi = 0. \tag{2.50}$$

Introducing the abbreviation $k = (2mW)^{1/2}/\hbar$, already used in Equation 2.47, and

# The calculation of permeabilities

making the substitutions

$$\psi = w\,\text{sech}^s\left(\frac{x}{2b}\right), \quad s = \frac{1}{2}\left[-1 + \left(1 - \frac{32mb^2 V_0}{\hbar^2}\right)^{1/2}\right] \tag{2.51}$$

Equation 2.50 becomes

$$\frac{d^2w}{dx^2} - \frac{s}{b}\tanh\left(\frac{x}{2b}\right)\frac{dw}{dx} + \left(\frac{s^2}{4b^2} + k^2\right)w = 0. \tag{2.52}$$

(The value of $s$ in Equation 2.51 has been chosen so as to make the coefficient of $w$ in Equation 2.52 independent of $x$.) Finally, a change of variable to $\xi = \sinh^2(x/2b)$ converts Equation 2.52 into

$$\xi(1+\xi)\frac{d^2w}{d\xi^2} + [(1-s)\xi + \tfrac{1}{2}]\frac{dw}{d\xi} + (\tfrac{1}{4}s^2 + b^2 k^2)w = 0. \tag{2.53}$$

Equation 2.53 has the same form as the standard hypergeometric equation, Equation A.2 of Appendix A, and its general solution in terms of hypergeometric functions is given by Equation A.3. The general expression for wave functions which satisfy the Schrödinger equation, Equation 2.50, is therefore

$$\psi = \text{sech}^s(x/2b)\{A_1 F[-\tfrac{1}{2}s + ikb, -\tfrac{1}{2}s - ikb, \tfrac{1}{2}, -\sinh^2(x/2b)]$$
$$+ A_2 \sinh(x/2b) F[-\tfrac{1}{2}s + ikb + \tfrac{1}{2}, -\tfrac{1}{2}s - ikb + \tfrac{1}{2}, \tfrac{3}{2}, -\sinh^2(x/2b)]\} \tag{2.54}$$

where $A_1$ and $A_2$ are arbitrary constants and $F$ denotes a hypergeometric function.

In order to calculate the barrier permeability it is necessary to examine the behaviour of Equation 2.54 as $x$ goes to $-\infty$ or $+\infty$. (cf. Fig. 2.7). This is done by using Equation A.4 of the Appendix to express Equation 2.54 in terms of hypergeometric functions of the variable $-1/\sinh^2(x/2b)$, which goes to zero for $x \to \pm\infty$. The hypergeometric functions on the right-hand side of Equation A.4 can then be replaced by unity, the first term of the expanison in Equation A.1. It is then readily shown that in both these extreme regions the wave function, Equation 2.54, has the asymptotic form $B_1 e^{ikx} + B_2 e^{-ikx}$, which is, of course, correct for a region of constant potential energy. The physical conditions of the problem demand that $B_2 = 0$ when $x \to +\infty$ (no particles approaching the barrier from the right), and this condition serves to determine the ratio of the coefficients $A_1$ and $A_2$ in Equation 2.54. Finally, the permeability is found as the ratio of the values of $|B_1|^2$ in the regions $x \to +\infty$ and $x \to -\infty$, and after a good deal of manipulation Equation 2.49 is obtained. A similar procedure can be used to derive Equation 2.48 for an unsymmetrical barrier.

## 2.4.2 Triangular or wedge-shaped barriers

Examples of this type of barrier are illustrated in Fig. 2.8. They are characterized by one or more sections in which the potential energy varies linearly with the distance. This will be the case for a charged particle moving in a uniform electric field, and Fig. 2.8a represents the example already mentioned (Section 1.4 and Fig. 1.3) of the field-assisted emission of electrons from a metal. In chemical problems a potential barrier is sometimes visualized in terms of a pair of intersecting potential energy curves, as in Fig. 2.9. For

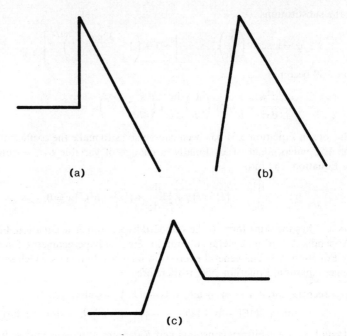

Fig. 2.8 Wedge-shaped and triangular barriers

moderate degrees of tunnelling the barrier can be approximated by a pair of intersecting straight lines, as shown by the region within the broken circle in the figure, and the problem can then be treated in terms of the infinite or truncated triangular barriers shown in Figs. 2.8b and c, respectively.

The calculation of the permeability in all these cases involves the solution of Schrödinger equations of the form

$$\frac{d^2\psi}{dx^2} + \frac{2m}{\hbar^2}(W - V_0 - Ax)\psi = 0 \qquad (2.55)$$

where $V_0$ is the potential energy at the apex of the wedge and $A$ is the slope of one of its sides, which may be either positive or negative. The equation is reduced to a simpler form by the substitution

$$\xi = \left(\frac{2mA}{\hbar^2}\right)^{1/3}\left(x + \frac{V_0 - W}{A}\right) \qquad (2.56)$$

when it becomes

$$\frac{d^2\psi}{d\xi^2} + \xi\psi = 0. \qquad (2.57)$$

If we now make the substitution

$$\psi = \xi^{1/2} Z(y), \quad y = \tfrac{2}{3}\xi^{3/2} \qquad (2.58)$$

# The calculation of permeabilities

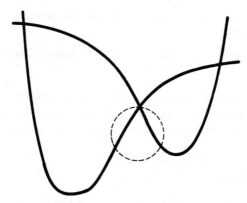

Fig. 2.9 Barrier formed by the intersection of two potential energy curves

it is easily shown that the function $Z(y)$ satisfies the equation

$$\frac{d^2 Z}{dy^2} + \frac{1}{y}\frac{dZ}{dy} + \left(1 - \frac{1}{9y^2}\right)Z = 0 \quad (2.59)$$

Equation 2.59 is Bessel's equation

$$\frac{d^2 Z}{dy^2} + \frac{1}{y}\frac{dZ}{dy} + \left(1 - \frac{n^2}{y^2}\right)Z = 0 \quad (2.60)$$

with $n = \pm \frac{1}{3}$, and a solution of Equations 2.55 and 2.57 is therefore

$$\psi = \xi^{1/2} Z_{\pm 1/3}(\tfrac{2}{3}\xi^{3/2}) \quad (2.61)$$

where $Z_{\pm 1/3}$ is any Bessel function of order $\pm 1/3$: the general solution will be a linear combination of any two such independent functions. The properties of Bessel functions and their derivatives are well known, and their values have been tabulated, so that the problem of calculating permeabilities for this class of barriers is in principle solved.

The calculations for the different types of barrier shown in Fig. 2.8 follow the methods described in the last three sections. Where there is a discontinuity in the potential energy the wave functions and their first derivatives on the two sides of the barrier must be matched, and the solution on the right-hand side of the barrier must be chosen so as to represent an outgoing wave as $x \to +\infty$, since the physical problem assumes that no particles are incident on the barrier from the right. The resulting expressions for the permeability are rather complicated [46, 47] and will not be given here.* When the particle energy is well below the top of the barrier these expressions can be converted with good

---

* The permeability is usually expressed in terms of the Airy functions $\mathrm{Ai}(\xi)$ and $\mathrm{Bi}(\xi)$, which are integral representations of the solution of Equation 2.57 and are simply related to the Bessel functions of order $\pm \frac{1}{3}$.

accuracy to much simpler forms. Thus for the potential represented by Fig. 2.8a the result is

$$G \approx \frac{4(V_0 - W)^{1/2} W^{1/2}}{V_0} \exp\left[-\frac{4\sqrt{(2m)}(V_0 - W)^{3/2}}{3\hbar A}\right]. \tag{2.62}$$

This expression was first derived by Fowler and Nordheim [15] for the emission of electrons from metals in strong electric fields ($A$ being then proportional to the field strength), as mentioned in Section 1.4.

### 2.4.3 The double anharmonic barrier

This barrier is defined by the equation

$$V(x) = V_0\left[1 - \left(\frac{a}{x} - \frac{x}{a}\right)^2\right]. \tag{2.63}$$

The potential energy has a maximum value $V_0$ at $x = a$, and goes to $-\infty$ at $x = 0, \infty$. The energy levels for the corresponding potential well (with $+V_0$ replaced by $-V_0$ in Equation 2.63) are well known [47, 48], but the corresponding barrier problem has only recently been solved [49]. The wave functions can be expressed in terms of the confluent hypergeometric function (cf. Appendix A), and if we write $v = (2V_0/ma^2)^{1/2}/\pi$, $B = 8mV_0 a^2/\hbar^2$ the expression found for the permeability is

$$G = \left(1 + \exp\left\{\frac{V_0 - W}{h\nu} + \frac{1}{8\pi}[(B-1)^{1/2} - B^{1/2}]\right\}\right)^{-1} \tag{2.64}$$

analogous to Equation 2.41 for a parabolic barrier.

### 2.4.4 The inverted Morse potential

This barrier is defined by the equation

$$V(x) = V_0(2e^{x/b} - e^{2x/b}) \tag{2.65}$$

and is thus an inversion of the Morse potential used in treating molecular vibrations. It has a maximum $V = V_0$ at $x = 0$, and the constant $b$ is a measure of the width of the barrier. The energy levels and wave functions of the Morse oscillator are, of course, well known, but the corresponding barrier problem has only recently been treated [50]. The results is

$$G = \frac{1 - \exp(-4\pi\lambda)}{1 + \exp[2\pi(\mu - \lambda)]} \tag{2.66}$$

where

$$\mu = (2mV_0)^{1/2} b/\hbar, \quad \lambda = (2mW)^{1/2} b/\hbar.$$

### 2.4.5 The barrier $V = V_0\{1 - |(x/b)^n|\}$

The above equation represents a symmetrical potential barrier for all positive

# The calculation of permeabilities

values of $n$. If $n \leq 1$ there is an abrupt change of slope at the maximum, while for $n > 1$ the curve is continuous. Particular cases of this barrier already treated are $n = 2$ (parabolic, Section 2.3), $n = 1$ (triangular, Section 2.4.2) and $n = \infty$ (rectangular, Section 2.2), but for values of $n$ other than these the resulting Schrödinger equation does not permit analytical solutions for arbitrary values of the energy $W$. However, such solutions do exist in the particular case $W = V_0$ (energy of particle matching the top of the barrier).

If the barrier extends to infinity in both the positive and the negative directions the resulting expression for the permeability is

$$G_0 = \sin^2[\pi/(n+2)]. \tag{2.67}$$

More complicated expressions can be derived for the case in which the barrier is truncated on either side by regions of constant potential energy. An incomplete treatment of the problem was given many years ago [51] and a more complete derivation is given in Appendix B.

These findings have little direct bearing on physical or chemical problems, since they apply only when $W = V_0$. However, they are useful in judging the accuracy of approximate expressions for the permeability when $W$ is close to $V_0$, and have some intrinsic interest, especially since the expression (Equation 2.67) depends only on $n$, and is independent of the parameter $b$ and the mass of the particle. Thus all infinite triangular barriers have $G_0 = \frac{1}{3}$, and all infinite parabolic barriers $G_0 = \frac{1}{2}$, the latter agreeing with the result of a detailed treatment (Equation 2.39).

This simple relation does not apply strictly to the truncated barriers treated in Appendix B, but it is shown there that Equation 2.67 remains a good approximation for most barriers having dimensions appropriate to chemical problems. Moreover, any curve having a finite curvature at its maximum can be approximated by a parabola in this region, and one might therefore expect that the exact expressions for the permeability of such barriers at $W = V_0$ will approximate to $G_0 = \frac{1}{2}$ for suitable values of the parameters. This is borne out by examining some of the expressions already obtained, for example:

Eckart potential (Equation 2.49)

$$G_0 \approx \tfrac{1}{2} \quad \text{if} \quad \exp[8b\sqrt{(2mV_0)}/\pi\hbar] \gg 1$$

Double anharmonic potential (Equation 2.64)

$$G_0 \approx \tfrac{1}{2} \quad \text{if} \quad \exp[32\pi a\sqrt{(2mV_0)}/\hbar] \gg 1$$

Inverted Morse potential (Equation 2.66)

$$G_0 \approx \tfrac{1}{2} \quad \text{if} \quad \exp[4\pi b\sqrt{(2mV_0)}/\hbar] \gg 1.$$

These conditions are all very similar and will be satisfied in many problems of chemical interest.

The derivation in Appendix B shows that Equation 2.67 (which does not involve the particle mass, the barrier dimensions, or Planck's constant) will be valid for any barrier if its height or width, or the particle mass, become sufficiently large. It will therefore apply to macroscopic systems: these would normally be treated by classical mechanics, which, however, make no prediction for the probability of motion across the barrier when $W$ is exactly equal to $V_0$.

### 2.4.6 The numerical calculation of barrier permeabilities

The potential barriers occurring in actual chemical or physical problems will not in general conform to any simple analytical function, and even when such a function is a good approximation to reality explicit expressions for the permeability can only be obtained for the few classes of function described in the preceding sections. There is thus ample scope for computations which involve the numerical integration of the Schrödinger equation over the relevant range of the co-ordinate $x$. This is particularly convenient when the barrier has regions of constant potential energy as $x \to \pm \infty$, in which the wave function is known to have the form $A_1 e^{ikx} + A_2 e^{-ikx}$. The remainder of the barrier is then divided into a large number of small sections, and the numerical procedure involves essentially the solution of the conditions of continuity of $\psi$ and $d\psi/dx$ between adjacent sections. Suitable procedures and computer programs have been worked out by R. J. LeRoy and his collaborators [52–54].

Such numerical calculations are obviously appropriate when information is available about the actual form of a particular energy surface. So far such information has only been obtained for simple gas-phase reactions such as $H + H_2$, and in general it may be preferable to choose an analytical barrier such as the Eckart function or the parabola and to vary its parameters so as to obtain the best fit with experiment. However, there are some suitable analytical barriers for which no explicit permeability expressions can be obtained, notably the *Gaussian barrier*, defined by

$$V(x) = V_0 \exp(-x^2/b^2) \tag{2.68}$$

which has been widely used. Because of the speed of modern computers the effect upon the permeability of varying the barrier parameters can be investigated numerically with little more trouble than when an explicit expression is available.

## 2.5 The BWK or semi-classical approximation

This approximate method of calculating permeabilities is applicable in principle to any type of one-dimensional barrier, and is particularly valuable when the permeability is low. It rests on a method of solving the Schrödinger equation proposed almost simultaneously by Brillouin [55, 56] Wentzel [57]

# The calculation of permeabilities

and Kramers [58] who used a general procedure originally due to Jeffreys [59]: hence the term BWK, WKB or JWKB approximation. The BWK approximate wave functions can be used either for determining the energy levels of a bound system, when they lead to the same results as the Bohr quantization rule, or to the problem of barrier penetration. The method is described in most textbooks on quantum theory, especially in the former context. However, a rigorous derivation involves a number of mathematical subtleties, particularly in the barrier problem, and the account given here makes no pretence at mathematical rigour. The problem has been considered in detail by a number of authors [60–63].

The one-dimensional Schrödinger equation can be written in the form

$$\frac{d^2\psi}{dx^2} + F(x)\psi = 0 \qquad (2.69)$$

where

$$F(x) = 2m[W - V(x)]/\hbar^2. \qquad (2.70)$$

If $F(x)$ is a constant the solution is simply $\psi = \exp(\pm iF^{1/2}x)$, or $\psi = \exp(\pm i\int_0^x F^{1/2}dx)$. This suggests that solutions of a similar form may be valid even when $F$ depends on $x$. If we write

$$\psi = \exp[if(x)] \qquad (2.71)$$

substitution in Equation 2.69 shows that $f(x)$ must satisfy the equation

$$i\frac{d^2f}{dx^2} - \left(\frac{df}{dx}\right)^2 + F = 0. \qquad (2.72)$$

The assumption made in the BWK approximation is

$$\left|\frac{d^2f}{dx^2}\right| \ll \left|\left(\frac{df}{dx}\right)^2\right| \qquad (2.73)$$

so that the first term of Equation 2.72 can be neglected, and direct integration gives

$$f = \pm \int_{x_0}^{x} F^{1/2} dx \qquad (2.74)$$

where the fixed limit of integration $x_0$ affects only an additive constant in $f$ and therefore a constant factor in $\psi$. When this is inserted in Equation 2.73 the condition for the validity of the approximation becomes

$$\left|\frac{d}{dx}(F^{1/2})\right| \ll |F| \quad \text{or} \quad \left|\frac{d}{dx}(F^{-1/2})\right| \ll 1. \qquad (2.75)$$

A further approximation is obtained by replacing the first term of Equation

2.72 by the value derived from Equation 2.74, which gives

$$\left(\frac{df}{dx}\right)^2 = F \pm i\frac{d}{dx}(F^{1/2}). \tag{2.76}$$

In taking the square root of Equation 2.76 we may apply the condition in Equation 2.75, so that

$$\frac{df}{dx} = \pm F^{1/2} + \frac{1}{2}iF^{-1/2}\frac{d}{dx}(F^{1/2}) \tag{2.77}$$

which gives on integration

$$f = \pm \int_{x_0}^{x} F^{1/2} dx + \tfrac{1}{2} i \ln F^{1/2} \tag{2.78}$$

and hence from Equation 2.71

$$\psi = F^{-1/4} \exp\left(\pm i \int_{x_0}^{x} F^{1/2} dx\right). \tag{2.79}$$

The solutions are of course valid only in regions for which Equation 2.73 or Equation 2.75 is satisfied, and the general solution will be a linear combination of the functions with positive and negative signs in the exponent.

Since the momentum $p_x$ is $[2m(W-V)]^{1/2}$, the integral occurring in Equation 2.79 is equivalent to $\hbar^{-1} \int p_x dx$, where $\int p_x dx$ is the *action*, a quantity which plays a prominent part in classical mechanics and in the Bohr theory of stationary states. The de Broglie wavelength of a particle is given by $\lambda = h/p_x = 2\pi\hbar/p_x$ (cf. Section 1.1), and the condition in Equation 2.75 is therefore equivalent to $|d(\lambda/2\pi)/dx| \ll 1$, i.e. the de Broglie wavelength must not vary appreciably over distances of the order of itself. Classical mechanics represents the limiting case in which the wavelength has become infinitely small, which explains the use of the term *semi-classical* (or *quasi-classical*) for the type of approximation described here.

An alternative (but equivalent) way of considering the problem is to write

$$\psi = \exp\left\{\frac{i}{\hbar}\int_{x_0}^{x}\left[f_0 + \frac{\hbar}{i}f_1 + \left(\frac{\hbar}{i}\right)^2 f_2 + \cdots \right]dx\right\}.$$

Since classical mechanics corresponds to $\hbar = 0$, successive approximations should be obtainable by breaking off the series after a given number of terms. By substituting in the Schrödinger equation and equating to zero the coefficients of successive powers of $\hbar$ we find

$$f_0 = [2m(W-V)]^{1/2}, \quad f_1 = \tfrac{1}{4}(W-V)^{-1}\frac{dV}{dx}, \quad \text{etc.}$$

The first approximation agrees with the result already obtained. Higher approximations have rarely been used in barrier problems, though they have been applied to calculate the energy levels of anharmonic oscillators [64]. However, in neither case can the accuracy of the result be improved indefinitely by this procedure, since the series diverges again if too many terms are taken.

# The calculation of permeabilities

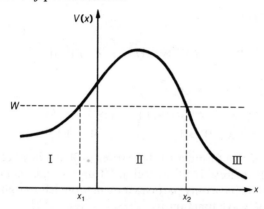

Fig. 2.10  Turning points and regions to be connected in the BWK treatment of barrier problems

Turning to the problem of barrier permeability (Fig. 2.10), it is clear that the wave function (Equation 2.79) applies only in regions I and III, where $W > V$, since $F$ has been treated as a positive quantity. For the non-classical region II, in which $W < V$, similar arguments lead to wave functions of the form*

$$\psi = (-F)^{-1/4} \exp\left[\pm \int_{x_0}^{x} (-F)^{1/2} dx\right] \quad (2.80)$$

again subject to the conditions in Equation 2.73 or Equation 2.75. The difficulty in solving the barrier problem is how to connect these functions at the points $x = x_1$ and $x = x_2$, since both wave functions become invalid in the neighbourhood of these points. This is readily seen from Equation 2.75, since $F^{-1/2}$ changes from $-\infty$ to $+\infty$ on passing through either of these points.†

The necessary connection formulae can only be derived by considering $x$ as a complex variable. The arguments resemble those used in Section 2.3 for the parabolic barrier, but are more complicated, since for other types of barrier the presence of complex zeros of $V(x)$ demands considerable care in choosing a satisfactory path of integration to join the two regions [60–63]. We shall give only the result here.

Since the limit of integration $x_0$ in Equations 2.79 and 2.80 is arbitrary, we may take it to be a point at which $W = V$ and hence $F = 0$. If we now define

---

* Since $(-1)^{1/2} = \pm i$ and $(-1)^{1/4} = \exp(\frac{1}{2}n + \frac{1}{4})\pi i$, where $n$ is an integer, Equations 2.79 and 2.80 are formally identical apart from a constant multiplying factor. However, since fractional powers of negative quantities can have several values, it is more satisfactory to use expressions in which only positive quantities are raised to fractional powers.

† Such points are frequently called *turning points*, since in classical mechanics they represent points at which the particle changes its direction of motion. The momentum at such points becomes zero and the de Broglie wavelength infinite, obviously contradicting the condition that this wavelength must not change appreciably over distances of the order of itself.

quantities $P$ and $Q$ by

$$P = \int_x^{x_0} F^{1/2} dx, \quad Q = \int_{x_0}^x (-F)^{1/2} dx \qquad (2.81)$$

the required connection formula is

$$F^{-1/4} e^{\pm iP \pm \pi i/4} \leftrightarrow (-F)^{-1/4}(e^Q \pm \tfrac{1}{2} i e^{-Q}). \qquad (2.82)$$
$$(x < x_0, W > V) \qquad (x > x_0, W < V)$$

The application of this formula to the barrier problem has been given in a simple form by B. Jeffreys [65]. In region III on the right-hand side of the barrier (Fig. 2.10) we have only particles moving from left to right, so that the appropriate BWK wave function is

$$\psi_{III} = F^{-1/4} \exp\left(i \int_{x_2}^x F^{1/2} dx\right) \qquad (2.83)$$

as may be seen by considering the special case in which $F$ is a constant. The connection formula (Equation 2.82) shows that the wave function in region II which connects with $\psi_{III}$ is

$$\psi_{III} = e^{-\pi i/4}(-F)^{-1/4} \left\{ \exp\left[\int_x^{x_2} (-F)^{1/2} dx\right] \right.$$
$$\left. + \tfrac{1}{2} i \exp\left[-\int_x^{x_2} (-F)^{1/2} dx\right] \right\}. \qquad (2.84)$$

If we now introduce the constant quantity $R$, defined in region II by

$$R = \int_{x_1}^{x_2} (-F)^{1/2} dx = \int_{x_1}^x (-F)^{1/2} dx + \int_x^{x_2} (-F)^{1/2} dx$$

Equation 2.84 becomes

$$\psi_{II} = e^{-\pi i/4}(-F)^{-1/4} \left\{ e^R \exp\left[-\int_{x_1}^x (-F)^{1/2} dx\right] \right.$$
$$\left. + \tfrac{1}{2} i e^{-R} \exp\left[\int_{x_1}^x (-F)^{1/2} dx\right] \right\}. \qquad (2.85)$$

The BWK wave function in region I will be of the form

$$\psi_I = A F^{-1/4} \exp\left[i \int_x^{x_1} F^{1/2} dx\right] + B F^{-1/4} \exp\left[-i \int_x^{x_1} F^{1/2} dx\right] \qquad (2.86)$$

in which the first and second terms represent the reflected and incident waves respectively, since $x$ is the lower limit of the integral. By the connection

## The calculation of permeabilities

formula (Equation 2.82) this connects with a function in region II given by

$$\psi_{\text{II}} = e^{-\pi i/4}(-F)^{-1/4}\left\{(A+iB)\exp\left[\int_{x_1}^{x}(-F)^{1/2}dx\right]\right.$$

$$\left. + \tfrac{1}{2}i(A-iB)\exp\left[-\int_{x_1}^{x}(-F)^{1/2}dx\right]\right\}. \quad (2.87)$$

Comparison of Equations 2.85 and 2.87 shows that $A = -i(e^R - \tfrac{1}{4}e^{-R})$, $B = e^R + \tfrac{1}{4}e^{-R}$. The permeability is given by

$$G = 1 - \frac{|A|^2}{|B|^2} \quad (2.88)$$

and hence

$$G = (e^R + \tfrac{1}{4}e^{-R})^{-2}. \quad (2.89)$$

If $e^{-R} \ll e^R$ this becomes

$$G = e^{-2R} = \exp\left[-\frac{2\sqrt{(2m)}}{\hbar}\int_{x_1}^{x_2}(V-W)^{1/2}dx\right] \quad (2.90)$$

which is the expression usually quoted for the barrier permeability in the BWK approximation.

We shall now note briefly the application of Equation 2.90 to some of the barriers already treated. For a *rectangular barrier* $F$ is a constant between $x_1$ and $x_2$, and $x_2 - x_1 = a$, the width of the barrier, so that

$$G = \exp\{-2[2m(V_0 - W)]^{1/2}a/\hbar\}. \quad (2.91)$$

This is identical with the exponential factor in Equation 2.24, though the pre-exponential factor (which may differ considerably from unity) is not obtained. For the *wedge-shaped barrier* of Fig. 2.8a and Section 2.4.2 we have $x_1 = 0$, $x_2 = (V_0 - W)/A$ (where $A$ is the slope of the wedge), so that the permeability is given by

$$G = \exp\left[-\frac{2\sqrt{(2m)}}{\hbar}\int_0^{(V_0-W)/A}(V_0 - Ax - W)^{1/2}dx\right]$$

$$= \exp\left[-\frac{4\sqrt{(2m)}}{3\hbar A}(V_0 - W)^{3/2}\right]. \quad (2.92)$$

This again agrees with the exponential part of the more exact expression (Equation 2.62), but omits the pre-exponential factor.

For the rectangular and wedge-shaped barriers the error caused by the absence of the pre-exponential factor from the BWK expressions arises from the abrupt changes in $V(x)$ which characterize these barriers, thus invalidating the connection formula, Equation 2.82, the derivation of which assumes that $V(x)$ is effectively linear and of finite slope in the neighbourhood of a turning point.* For the less artificial 'well-rounded' barriers which are more appropriate to chemical problems the agreement of the BWK result with the exact value is much better. Thus for the *infinite parabolic barrier* $V(x) = V_0 - \frac{1}{2}Ax^2$ the BWK expression is

$$G = \exp\left\{-\frac{2\sqrt{(2m)}}{\hbar}\int_{-[2(V_0-W)/A]^{1/2}}^{+[2(V_0-W)/A]^{1/2}}(V_0 - \tfrac{1}{2}Ax^2 - W)^{1/2}dx\right\}$$
$$= \exp\left[-\frac{2\pi m^{1/2}(V_0 - W)}{\hbar A^{1/2}}\right]. \tag{2.93}$$

which agrees with the exact value, Equation 2.39, provided that the permeability is small.

The evaluation of the integral in Equation 2.90 can often be simplified by using the following general result, which can be derived by standard methods. If $a + by + cy^2$ is any quadratic expression in which $a < 0$, $c < 0$ and $b^2 - 4ac > 0$, and if $y_1$ and $y_2$ are the roots of $a + by + cy^2 = 0$ ($y_2 > y_1$), then

$$\int_{y_1}^{y_2} \frac{(a + by + cy^2)^{1/2}\,dy}{y} = \frac{1}{2}\pi\left[\frac{b}{(-c)^{1/2}} - 2(-a)^{1/2}\right]. \tag{2.94}$$

By applying Equation 2.94 to the symmetrical *Eckart barrier* (Section 2.4.1), after making the substitution $y = \exp(x/b)$, we find

$$G = \exp[-4\pi\sqrt{(2m)}b(V_0^{1/2} - W^{1/2})/\hbar] \tag{2.95}$$

while the corresponding expression for the unsymmetrical barrier defined by Equation 2.45 is

$$G = \exp\{-2\pi\sqrt{(2m)}b[A^{1/2} - W^{1/2} - (W - B)^{1/2}]/\hbar\}. \tag{2.96}$$

For the *inverted Morse potential* (Section 2.4.4) the same substitution yields the result

$$G = \exp[-2\pi\sqrt{(2m)}b(V_0^{1/2} - W^{1/2})/\hbar] \tag{2.97}$$

while for the *double anharmonic potential* (Equation 2.4.3) the substitution $y = x^2/a^2$ gives

$$G = \exp\left[-\frac{\pi\sqrt{(2m)}a}{2\hbar}\left(\frac{V_0 - W}{V_0^{1/2}}\right)\right]. \tag{2.98}$$

Each of the last four equations agrees with the corresponding exact expression

---

* This difficulty does not arise in applying the BWK method to triangular barriers such as Figs. 2.8b and 2.8c, since the potential energy curve no longer has a vertical section.

# The calculation of permeabilities

when the permeability is much less than unity, i.e. when the particle energy is appreciably lower than the top of the barrier.

The number of barrier types for which the BWK integral can be explicitly integrated is strictly limited, though since only a simple integration is involved Equation 2.90 can be rapidly evaluated numerically for a barrier of arbitrary form. In addition to the examples given above, Christov [66] has derived explicit expressions for the BWK permeability of a generalization of the Eckart potential (Equation 2.45), which contains one further parameter in addition to $A$, $B$, and $b$: it represents a family of barriers of constant width and height, but variable curvature, and includes as special cases the Eckart potential and the truncated parabola. For some chemical problems a reasonable type of barrier is $V(x) = \cos^2(\pi x/2b)$ between $x = -b$ and $x = +b$. The BWK permeability in this case can be expressed in terms of complete elliptic integrals [67], though this result does not appear to have been applied to chemical problems. The wave functions for this barrier involve Mathieu functions and an exact expression can be obtained for the permeability, though it is complicated and unsuitable for computation [68].

Equations 2.89 and 2.90 and any expressions derived from them will be good approximations when $W$ is considerably less than $V_0$ and the permeability is small. They may be expected to fail when $W$ is close to $V_0$, since there is then no part of region II in which the condition in Equation 2.75 is satisfied. The maximum error will arise when $W = V_0$. The values of $G$ predicted by Equations 2.89 and 2.90 are then respectively 16/25 and unity, while it has been shown in Section 2.4.5 that the true value is close to one half for any barrier having a finite radius of curvature at the top. Moreover, the procedure given above cannot be used to treat the reflection of particles from the barrier when $W > V_0$.

These deficiencies are largely removed by a more thorough treatment of the BWK problem [61]. Instead of Equation 2.89 or Equation 2.90 the expression obtained is

$$G = (1 + e^{2R})^{-1}$$
$$= \left\{ 1 + \exp\left[\frac{2\sqrt{(2m)}}{\hbar} \int_{x_1}^{x_2} (V - W)^{1/2} dx\right] \right\}^{-1}. \quad (2.99)$$

This expression reduces to Equation 2.90 for low permeabilities, but it has the correct value $G = \frac{1}{2}$ when $W = V_0$ and is identical with the exact result for a parabolic barrier (Equation 2.39). The same procedure yields a simple expression for the permeability when $W > V_0$, and Equation 2.99 agrees better than Equation 2.90 with the exact results for the six other types of barrier treated in this section. However, this more accurate version of the BWK result has been rarely used, and it has been common practice to employ Equation 2.90 together with the assumption that $G = 1$ when $W > V_0$.

## 2.6 Tunnelling by bound particles

The situations envisaged in this type of calculation have been illustrated in Fig. 2.1b and c. Although this kind of model might appear particularly appropriate to chemical problems, which frequently involve the motion of particles to or from bound states, it has in fact been used much less than the free-particle model so far considered in this chapter. This arises from a number of causes. In the first place the concepts involved (such as weak quantization, complex eigenvalues and time-dependent perturbation theory) are less familiar to most chemists. Secondly, it is usually impossible to arrive at an exact result, and it is difficult to assess the errors involved in approximate treatments: in particular, the tunnelling frequency in the double-minimum problem (Fig. 2.1c) turns out to be very sensitive to slight departures of the system from symmetry. Finally, it remains doubtful whether the bound-particle model is really to be preferred for treating tunnelling in chemical reactions, as will be discussed in Chapter 7.

For these reasons only a superficial account of the theory is given in this section, though a more sophisticated treatment gives essentially the same results.

### 2.6.1 *Systems with a single minimum*
The type of potential energy curve concerned is shown in Fig. 2.11. It has been used particularly as a model for radioactive disintegration, but it will also apply to the dissociation of a molecule into two parts, provided that this process involves energy of activation.

The simplest treatment involves a mixture of classical and quantum-

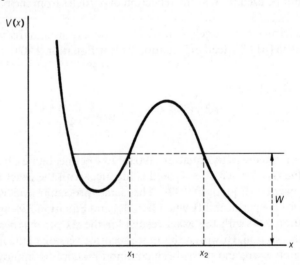

Fig. 2.11   Tunnelling in a system with a single potential energy minimum

# The calculation of permeabilities

mechanical concepts, and was originally given by von Laue [69]. According to classical mechanics a particle inside the potential well will vibrate between $x_1$ and $x_2$ with a frequency $v$, which is in general a function of the energy $W$, while quantum theory predicts that on each impact with the barrier there is a probability $P$ of escape, which is also a function of $W$. If we consider a number of bound systems, all of energy $W$, the proportion which 'dissociate' in unit time is $Pv$, which therefore represents the first-order velocity constant for the dissociation of systems of this energy. In particular, if the BWK expression (Equation 2.90) for permeability to free particles is valid here this first-order constant is given by

$$\lambda = v\exp\left[-\frac{2\sqrt{(2m)}}{\hbar}\int_{x_1}^{x_2}(V-W)^{1/2}\,dx\right]. \tag{2.100}$$

Because of the unpalatable mixture of classical and quantal arguments, this derivation is not very satisfactory. Since the potential energy does not become infinite on the right-hand side of the diagram there are no exactly defined energy levels within the barrier, a state of affairs usually described as *weak quantization*. However, essentially the same result is obtained by a closer examination of the problem [70], and we shall not pursue the matter further here.

## 2.6.2 Symmetrical double-minimum potentials

We shall first consider the completely symmetrical situation illustrated in Fig. 2.12 in which both the heights and the shapes of the energy minima are

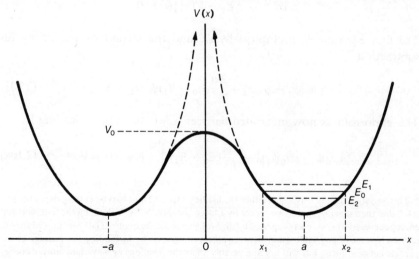

Fig. 2.12 Splitting of energy levels in a double-minimum potential

identical on the two sides of the barrier. This problem was first considered by Dennison and Uhlenbeck [71] in connection with the doubling of lines in the vibrational spectrum of ammonia, and most of its subsequent applications have been to spectroscopy rather than to chemical reactions, as outlined in Section 6.3.

If we consider one of the potential wells in isolation* there will be a series of non-degenerate energy levels, of which we shall single out one, of energy $E_0$ and normalized wave function $\psi_0$. However, for a system consisting of two such potential wells each level becomes doubly degenerate, since there are now two wave functions, which to a first approximation can be written

$$\psi_1(x) = 2^{-1/2}[\psi_0(x) + \psi_0(-x)]$$
$$\psi_2(x) = 2^{-1/2}[\psi_0(x) - \psi_0(-x)] \quad (2.101)$$

where the origin is halfway between the minima of the two wells, so that $\psi_0(x)$ and $\psi_0(-x)$ refer respectively to the right-hand and left-hand wells. The factor $2^{-1/2}$ implies that normalization now extends over both wells. If the overlap between $\psi_0(x)$ and $\psi_0(-x)$ can be neglected both wave functions lead to the same energy $E_0$, but if there is appreciable overlap they predict different energies $E_1$ and $E_2$, with $E_1 < E_0 < E_2$, as in the analogous LCAO treatment of the hydrogen molecule ion. Provided that the splitting $E_2 - E_1$ is small a general expression for its magnitude is easily obtained.

The Schrödinger equations for $\psi_0$ and $\psi_1$ are respectively

$$\psi_0'' + \frac{2m}{\hbar^2}[E_0 - V(x)]\psi_0 = 0$$
$$\psi_1'' + \frac{2m}{\hbar^2}[E_1 - V(x)]\psi_1 = 0. \quad (2.102)$$

The first equation is multiplied by $\psi_1$ and the second by $\psi_0$, giving on subtraction

$$\psi_0''\psi_1 - \psi_0\psi_1'' + \frac{2m}{\hbar^2}(E_0 - E_1)\psi_0\psi_1 = 0. \quad (2.103)$$

This expression is now integrated between $x = 0$ and $x = \infty$, leading to

$$(\psi_0'\psi_1 - \psi_0\psi_1')_0^\infty + \frac{2m}{\hbar^2}(E_0 - E_1)\int_0^\infty \psi_0\psi_1 \, dx = 0. \quad (2.104)$$

---

* This 'isolation' can be envisaged either by dividing Fig. 2.12 into two by cutting down the line $x = 0$ and then separating the two halves by a large distance, or by prolonging the two potential energy curves to $V(x) = \infty$ as suggested by the broken lines in the Figure. The latter procedure will effectively remove any interaction between the two systems by forbidding the passage of the particle between them, but will not appreciably affect the positions or wave functions of energy levels which lie well below the maximum of potential energy.

# The calculation of permeabilities

The wave functions and their derivatives tend to zero as $x \to \infty$, while at $x = 0$ $\psi_1(0) = 2^{1/2}\psi_0(0)$ and $\psi'_1(0) = 0$. Moreover, since $\psi_0(-x)$ diminishes exponentially outside the left-hand well, it is a good approximation to write

$$\int_0^\infty \psi_0\psi_1\,dx \approx 2^{-1/2} \int_0^\infty \psi_0^2\,dx = 2^{-1/2}$$

giving finally

$$E_1 - E_0 = -\frac{\hbar^2}{m}\psi_0(0)\psi'_0(0). \qquad (2.105)$$

A similar calculation gives for $E_2 - E_0$ the same expression with the opposite sign, so that we have for the splitting of the two levels

$$\Delta E = E_2 - E_1 = \frac{2\hbar^2}{m}\psi_0(0)\psi'_0(0). \qquad (2.106)$$

Further progress is made by introducing the BWK approximation for $\psi_0$. Within the right-hand well, where $E > V$, the form of the solution is given by Equation 2.79 as

$$\psi = F^{-1/4}\exp\left(\pm\int_0^x F^{1/2}\,dx\right) \qquad (2.107)$$

where $F = 2m[E - V(x)]/\hbar^2$. In the present problem we are dealing with a stationary state, with no nett flux of particles, so that the coefficients of the terms with positive and negative exponents must have the same absolute value (cf. Equation 2.5). In view of the connection formula, Equation 2.82, it is convenient to write

$$\psi_0 = CF^{-1/4}\sin\left[\int_0^x F^{1/2}\,dx + \frac{1}{4}\pi\right] \qquad (2.108)$$

where $C$ is the normalization factor. Since the wave function decays exponentially outside the classical limits of motion $x = x_1$ and $x = x_2$ (see Fig. 2.12), it is sufficiently accurate to use these limits in determining $C$, i.e. we write

$$\int_{x_1}^{x_2}\psi_0^2\,dx = C^2\int_{x_1}^{x_2} F^{-1/2}\sin^2\left(\int F^{1/2}\,dx + \frac{1}{4}\pi\right)dx = 1. \qquad (2.109)$$

If the interval of integration contains a sufficiently large number of wavelengths of the oscillatory function we can replace the $\sin^2$ term by its average value, $\frac{1}{2}$, giving

$$\frac{1}{2}C^2\int_{x_1}^{x_2} F^{-1/2}\,dx = 1. \qquad (2.110)$$

Since the function $F$ can be written as $m^2v^2/\hbar^2$, where $v$ is the particle velocity, the integral in Equation 2.110 is equivalent to

$$\frac{\hbar}{m}\int_{x_1}^{x_2} dx \bigg/ \frac{dx}{dt} = \frac{\hbar}{2mv} \qquad (2.111)$$

where $v$ is the classical vibration frequency, which is in general a function of the energy. Equation 2.110 then gives $C = (4mv/\hbar)^{1/2}$, and the final expression for $\psi_0$ within the potential well is

$$\psi_0 = (4mv/\hbar)^{1/2} F^{-1/4} \sin\left(\int_0^x F^{1/2} dx + \frac{1}{4}\pi\right). \qquad (2.112)$$

In order to evaluate Equation 2.106 we need an expression for $\psi_0$ inside the barrier, where $E < V$. Since $\sin\theta = -\frac{1}{2}i(e^{i\theta} - e^{-i\theta})$, the connection formula (Equation 2.82) gives in this region

$$\psi_0 = (mv/\hbar)^{1/2}(-F)^{-1/4} \exp\left[-\int_x^{x_1} (-F)^{1/2} dx\right] \qquad (2.113)$$

whence

$$\psi_0(0) = \frac{(mv)^{1/2}}{[2m(V_0 - E_0)]^{1/4}} \exp\left[-\frac{\sqrt{(2m)}}{\hbar}\int_0^{x_1} (V - E_0)^{1/2} dx\right]$$

$$\psi_0'(0) = \frac{[2m(V_0 - E_0)]^{1/2}}{\hbar}\psi_0(0) \qquad (2.114)$$

and hence finally from Equation 2.106

$$\Delta E = E_2 - E_1 = 2\hbar v \exp\left[-\frac{2\sqrt{(2m)}}{\hbar}\int_0^{x_1}(V - E_0)^{1/2} dx\right]$$

$$= 2\hbar v \exp\left[-\frac{\sqrt{(2m)}}{\hbar}\int_{-x_1}^{+x_1}(V - E_0)^{1/2} dx\right] \qquad (2.115)$$

where the last step follows from the symmetry of the potential energy curve about $x = 0$.

As is often the case in this type of treatment, the validity of the above derivation requires careful examination. The use of the unperturbed wave functions (Equation 2.101) and the approximation made in arriving at Equation 2.105 imply that the situation differs only slightly from that which prevails when the two potential wells are isolated, i.e. the energy levels concerned must lie well below the top of the barrier. On the other hand, the BWK approximation is in general only valid for high quantum numbers, and is inaccurate for energy levels near the bottom of the potential well: this restriction applies in particular to the argument used to determine the normalization factor from Equation 2.110. [In other words, the intersections which the line $V(x) = E_0$ makes with the potential energy curve should be well separated *both* for the barrier *and* for the potential wells.] Since these two requirements oppose one another, the validity of the whole procedure might be doubted. However, a quantitative examination shows that for problems of chemical interest the majority of levels will satisfy both requirements with a high degree of accuracy. Moreover, in the particular case of a parabolic potential well (harmonic oscillator) the BWK treatment yields exact wave functions and eigenvalues even down to the lowest level, and we have already seen in Section 2.5 that the same treatment gives exact values for the permeability of a parabolic barrier. Since any continuous function with a finite curvature can be approximated by a parabola at its maxima and minima, Equation 2.115 can be applied with some confidence to a wide range of problems.

If the form of the potential energy curve permits exact solutions of the wave equation these can of course be used in place of the BWK approximation to derive the splitting of the energy levels, though this procedure is rarely useful in practice. Thus for a highly artificial system consisting of two rectangular potential wells separated by a rectangular barrier there is no difficulty in writing down the exact wave functions, but no simple expression results for the splitting of the doublet. A more realistic function for which exact wave functions are available is that treated by Manning [72], $V(x) = A \operatorname{sech}^4(x/2\rho) - B \operatorname{sech}^2(x/2\rho)$, where $A$, $B$, and $\rho$ determine the shape and dimensions of the function, but here again the doublet splitting is not obtained directly. It has therefore been usual to employ the BWK treatment or some other type of approximation. As we have seen in connection with free-particle tunnelling (Section 2.5), the integral in Equation 2.115 can be evaluated analytically for a few types of barrier, but frequently must be calculated numerically. A useful approximate treatment, valid when $\Delta E$ is small and the wells are approximately parabolic, has been given by Harmony [73]. The result is

$$\Delta E = \frac{2ah^2\alpha^{3/2}}{m\pi^{1/2}} \exp(-\alpha a^2) \qquad (2.116)$$

in which the distance between the two minima is $2a$, and $\alpha = 2\pi v m/\hbar$, where $v$ is the frequency of oscillation in each well. It is interesting to note that, to this approximation, $\Delta E$ is independent of the height of the barrier, but depends strongly on both $a$ and $v$.

Up to this point our treatment of the double-minimum problem has made no mention of tunnelling, or indeed of any time-dependent phenomenon, apart from the vague implication that the splitting of a vibrational level is related to the ability of the particle to pass from one well to the other. The wave functions $\psi_1(x)$ and $\psi_2(x)$ (Equation 2.101) refer to *stationary states* of the system, and even if either of them is multiplied by the factor $\exp(-iEt/\hbar)$ so as to give solutions of the time-dependent Schrödinger equation (Equation 2.2) this factor disappears in the products such as $\Psi(x, t) \Psi^*(x, t)$ which occur in calculating the probability of observing certain values for physical quantities (e.g. the position of the particle), or the most probable values of such quantities. For example, if measurement shows that the particle has an energy $E_1$, then the standard interpretation of $\psi_1$ is that at the time of measurement there was an equal probability that the particle had $x > 0$ or $x < 0$, or, somewhat more loosely, that it had an equal probability of being in either well. The same is true of a particle of energy $E_2$ and wave function $\psi_2$.

The situation is different if we suppose that at a certain time $t = 0$ the particle is definitely observed to be in one of the wells, with an energy $E_0 \approx E_1 \approx E_2$. If this is the left-hand one the appropriate wave function is $\psi_0(-x)$, but this no longer represents a stationary state of the whole system, and we cannot construct the time-dependent wave function merely by multiplying by $\exp(-iE_0t/\hbar)$. Instead we must express $\psi_0(-x)$ in terms of the stationary state functions $\psi_1$ and $\psi_2$, giving from Equation 2.101

$$\Psi = \tfrac{1}{2}\sqrt{2}[\Psi_1(x,t) - \Psi_2(x,t)]$$
$$= \tfrac{1}{2}\sqrt{2}[\psi_1(x)e^{-iE_1t/\hbar} - \psi_2(x)e^{-iE_2t/\hbar}]. \qquad (2.117)$$

The probability distribution at any time $t$ is then given by $\Psi\Psi^*$, which is easily shown to be

$$\Psi\Psi^* = \tfrac{1}{2}[\psi_1^2(x) + \psi_2^2(x) - 2\psi_1(x)\psi_2(x)\cos(t\Delta E/\hbar)]$$
$$= \psi_0^2(x)\sin^2(\tfrac{1}{2}t\Delta E/\hbar) + \psi_0^2(-x)\cos^2(\tfrac{1}{2}t\Delta E/\hbar) \qquad (2.118)$$

where $\Delta E = E_2 - E_1$.

Equation 2.118 implies that the particle tunnels back and forth periodically between the two wells, the time taken to pass from one to the other being given by $t\Delta E/\hbar = \pi$, i.e. $t = h/2\Delta E$, where $h = 2\pi\hbar$. This corresponds to a *tunnelling frequency*\* $v_t$ given by

$$v_t = 2\Delta E/h \qquad (2.119)$$

which is the fundamental relation between the splitting of a vibrational energy level and the rate of the tunnelling process. More sophisticated treatments can be given [74, 75], but Equation 2.119 is satisfactory provided that the doublet splitting is small compared with the separation between states of different vibrational quantum numbers.

If we use the BWK expression (Equation 2.115) for $\Delta E$ we obtain for the tunnelling frequency

$$v_t = \frac{2v}{\pi}\exp\left[-\frac{\sqrt{(2m)}}{\hbar}\int_{-x_1}^{+x_1}(V-E_0)^{1/2}dx\right]. \qquad (2.120)$$

This expression bears a close resemblance to Equation 2.100, previously derived for the escape frequency of a particle from a *single* potential well on the basis of the BWK expression for free-particle tunnelling. There is, however, a very important difference, in that the exponent in Equation 2.100 contains a factor 2 which is absent in Equation 2.120. Since in both cases the value of the exponential is normally very much less than unity there is a large difference between the two values, and the application of Equation 2.100 to tunnelling

---

\* There is some ambiguity in the literature about the term 'tunnelling frequency', which is sometimes defined in terms of the complete cycle left-hand well → right-hand well → left-hand well, thus giving a frequency which is one half of that used here.

between two symmetrical potential wells will therefore lead to large errors. The physical basis for this difference lies in a kind of resonance between the two wells which is absent when the particle can escape to infinity in one or both directions [76].

The possibility of observing non-stationary states experimentally, and hence of making direct measurements of tunnelling frequencies, needs careful examination. In general terms the answer is that such observations are possible if the characteristic time involved in the measuring technique is small compared with $1/v_t$ [67]. This result is fairly obvious if we regard the transfer of the particle from one well to the other as a chemical reaction which is being monitored by measuring the concentrations of one or both species, though it should be noted that if the potential energy curve is truly symmetrical the two states are unlikely to show appreciable differences in chemical or physical properties (e.g. absorption spectra).* More subtle implications arise in considering the effect of tunnelling in molecular spectra, which will be discussed in Chapter 6.

### 2.6.3 Unsymmetrical double-minimum potentials

In the case of tunnelling by a free particle the permeability is little changed by passing from a symmetrical barrier to one which is slightly unsymmetrical.†
By contrast, for a double-minimum potential slight departures from symmetry always produce a large decrease in the tunnelling frequency, and at the same time render the problem much more difficult to treat theoretically.

Crudely speaking, the decrease in tunnelling frequency may be attributed to the fact that the introduction of asymmetry destroys the exact state of resonance between the two wells. The stationary state wave functions can no longer be classed as symmetric or antisymmetric, as in the symmetrical case (cf. Equation 2.101), and they no longer predict equal probabilities of finding the particles in the two wells. For example, if the right-hand well is slightly lower than the left-hand one the lowest stationary state $\Psi_0$ corresponds to a particle which is largely localized in the former, while for the next lowest state $\Psi_1$ this localization is transferred to the left-hand well. The non-stationary state in which the particle is definitely known to be in the left-hand well will be

---

* The most favourable technique from this point of view is electron spin resonance, since two isomeric radicals may have quite distinct e.s.r. spectra even when they have almost identical energies. Examples of this type of reaction are given in the review of experimental evidence in Chapter 5.

† It is sometimes stated that when considering the passage of a free particle from left to right across a barrier the permeability to a particle of given energy is increased if the potential energy on the right-hand side of the barrier is decreased, and vice versa. This statement needs qualification, since the sign of the effect may depend on the actual value of the permeability, and also on which barrier parameters (such as width or curvature at the maximum) are held constant on passing from the symmetrical to the unsymmetrical case. However, it remains true in general that for free particles only small effects are produced by small departures from symmetry.

approximately represented by $\Psi_1 + c\Psi_0$, where $c^2 \ll 1$, and the average time which elapses before the particle appears in the right-hand well will be greater than that in the symmetrical case by a factor of the order of $1/c^2$. A similar difference arises if the shapes of the two wells are slightly different, even if their minima have exactly the same energies.

A number of authors have treated in detail the problem of tunnelling between asymmetric double minima [75–79]. The problem is a complex one, and only some general conclusions will be mentioned here. The effect of asymmetry is always to reduce the tunnelling frequency, and an energy difference between the two minima amounting to a few per cent of the barrier height can lead to a reduction factor of several powers of ten when the tunnelling probability (defined as the ratio of the tunnelling frequency to the classical vibration frequency) is much less than unity. Harmony [80] has given the following approximate expression, valid when both the asymmetry and the tunnelling probability are small. If $v_t^0$ and $v_t$ refer respectively to the symmetrical and unsymmetrical systems and $\Delta V$ is the energy difference between the two minima in the latter, then

$$\frac{v_t}{v_t^0} = \frac{2hv_t^0}{[(hv_t^0)^2 + 16(\Delta V)^2]^{1/2}}. \tag{2.121}$$

The large quantitative differences between tunnelling in symmetrical and unsymmetrical double minima have led some authors [79] to suggest that the two phenomena have fundamentally different explanations, the former being attributed to the identity of the two potential wells and the latter to the uncertainty principle. However, this seems an unnecessarily extreme point of view. The uncertainty principle is fundamental to a large part of quantum theory, including all tunnelling phenomena, and it seems preferable to regard the increased tunnelling in the symmetrical case as due to a resonance enhancement which is analogous to the enhanced rate of energy transfer arising in classical mechanics when two vibrations have identical frequencies.

# 3 The application of tunnel corrections in chemical kinetics

## 3.1 Barrier permeabilities for a Boltzmann energy distribution

The preceding chapter deals with the permeability of one-dimensional barriers to particles of specified energy $W$. Under normal conditions a chemical reaction involves a large number of systems covering a range of energies, and in this context a more realistic model consists of a stream of particles in thermal equilibrium impinging on a barrier. The simplest form of energy distribution in such a stream is given by

$$\frac{dN}{N} = \frac{1}{kT} e^{-W/kT} dW \qquad (3.1)$$

in which $dN/N$ is the fraction of particles having energies between $W$ and $W + dW$.

This expression is exact only when the energy is expressible as the sum of two terms, each of which is proportional either to the square of a momentum (as in any form of kinetic energy) or to the square of a co-ordinate (as in the potential energy of a harmonic oscillator). It is therefore not strictly applicable to a stream of particles moving in one dimension, as envisaged here. However, two square terms are involved in many situations relevant to chemical kinetics, for example the total energy (kinetic plus potential) of a harmonic oscillator, or the relative kinetic energy along the line of the centres of two colliding particles. It is therefore reasonable to use the simple type of Boltzmann distribution given by Equation 3.1 in applications to chemical kinetics.

If $J_0$ is the total flux of particles striking the left-hand side of the barrier and $G(W)$ the permeability for an energy $W$, the rate $J$ at which particles appear on the right-hand side of the barrier is given by

$$J = \frac{J_0}{kT} \int_0^\infty G(W) e^{-W/kT} dW. \qquad (3.2)$$

If classical mechanics were obeyed we should have $G(W) = 0$ for $W < V_0$ and $G(W) = 1$ for $W > V_0$, where $V_0$ is the potential energy at the top of the barrier. The classical rate $J_c$ is therefore

$$J_c = \frac{J_0}{kT} \int_{V_0}^\infty e^{-W/kT} dW = J_0 e^{-V_0/kT}. \qquad (3.3)$$

By combining Equations 3.2 and 3.3 we can formulate a *tunnel correction* $Q_t$ which is the ratio of the quantum-mechanical rate to the classical rate, i.e.

$$Q_t = \frac{J}{J_c} = \frac{e^{V_0/kT}}{kT} \int_0^\infty G(W) e^{-W/kT} dW. \tag{3.4}$$

The integrand in Equation 3.4 contains two opposing factors, $\exp(-W/kT)$ which increases with decreasing $W$, and $G(W)$ which decreases with decreasing $W$. The product $G(W) \exp(-W/kT)$ represents the distribution of transmitted particles as a function of energy. We have seen in Section 2.4.5 that for well-rounded barriers of chemical interest $G(W)$ is approximately equal to one half when $W = V_0$, and goes asymptotically towards unity for increasing values of $W - V_0$. In this region, therefore, ('above the barrier') the dominant factor is $\exp(-W/kT)$ and the distribution curve will always approach the classical one with increasing $W$. In the region $W < V_0$ ('below the barrier') the form of the distribution curve depends upon the quantitative behaviour of $G(W)$ and hence on the shape and dimensions of the barrier and the mass of the particle. At one extreme of behaviour the decrease in $G(W)$ with decreasing $W$ may rapidly outweigh the increase in $\exp(-W/kT)$, so that very little contribution is made by particles with $W < V_0$, and the whole behaviour is close to classical. At the other extreme the increase in the exponential term may dominate over the decrease in $G(W)$ right down to $W = 0$; the distribution curve then has no maximum and the major contribution comes from systems with energies much less than $V_0$, corresponding to highly non-classical behaviour.

It seems unlikely that this last extreme state of affairs applies to chemical reactions at ordinary temperatures. In many instances, especially at high temperatures, behaviour may be almost classical, corresponding to the first extreme case mentioned above. However, we shall see later that for reactions involving isotopes of hydrogen there is often evidence for tunnel corrections of moderate magnitude, when the position is intermediate between the two extremes described. A typical distribution curve for this situation is shown in Fig. 3.1, together with the corresponding distribution function for the classical case. When the tunnel effect is taken into account there is a maximum in the distribution curve not far from $W = V_0$, and a considerable proportion of the transmitted particles have energies less than $V_0$. The value of the integral in Equation 3.4 is given by the area under the distribution curve, and the tunnel correction $Q_t$ is equal to the ratio of this area to the area under the classical curve.

If the particle energies form a discrete rather than a continuous set, then Equation 3.2 must be replaced by

$$J = \sum_i v_i G(W_i) \exp(-W_i/kT) \tag{3.5}$$

# The application of tunnel corrections

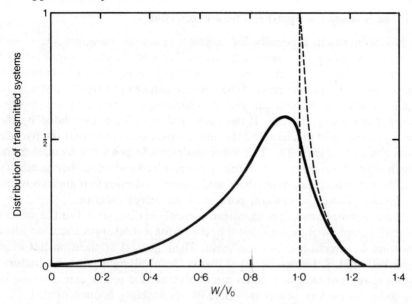

Fig. 3.1 Energy distribution of reacting systems in classical and quantum mechanics. Calculated for H-transfer at 298 K across a symmetrical parabolic barrier of height 41.8 kJ mol$^{-1}$ and width 93 pm

where $v_i$ is the frequency with which particles of energy $W_i$ strike the barrier. The frequencies and energy levels depend on the shape of the energy well which gives rise to the vibrational or other quantization, and it has been usual to retain the integration in Equations 3.2 and 3.4. The justification for this procedure will be examined in Chapter 7.

If the tunnelling correction factor $Q_t$ is close to unity, corresponding to small deviations from classical behaviour, it can be expressed in a simple form which is independent of the particular shape of the barrier, provided that it has a finite radius of curvature at the top. It was shown by Wigner [38], using very general arguments, that under these conditions the first-order tunnel correction is given by

$$Q_t = 1 - \frac{\hbar^2}{24mk^2T^2}\left(\frac{d^2V}{dx^2}\right)_0 = 1 + \frac{1}{24}\left(\frac{hv_\ddagger}{kT}\right)^2 \tag{3.6}$$

where $(d^2V/dx^2)_0$ refers to the top of the barrier and $iv_\ddagger$ is the imaginary barrier frequency as defined in Section 2.3 (Equation 2.40). It will be seen later that the general expression for $Q_t$ for a parabolic barrier reduces to Equation 3.6 for high temperatures or small values of $v_\ddagger$.

## 3.2 The principles of applying a tunnel correction

It has been usual, especially for solution reactions, to apply the tunnel correction simply by multiplying the classical or semi-classical* expression for the reaction velocity by the correction factor $Q_t$ as defined in Equation 3.4. This implies that the progress of the reaction can be modelled by a stream of particles with a Boltzmann energy distribution impinging on a one-dimensional energy barrier. If the permeability $G(W)$ is calculated by the methods described in Chapter 2 this implies the assumption that the particle mass has a constant value. This value represents in general a reduced mass, which depends on the masses of the atoms involved and upon their geometry and the forces acting between them, and which is assumed to remain constant for motion within the relevant portion of the energy diagram.

These assumptions will be examined critically in Chapter 7. For the present we shall accept the statement that they constitute a good approximation when the tunnel correction is not too great. Their general application has been described by H. S. Johnston [82] as "better chemical engineering than natural philosophy", but lack of a better alternative has led to the extensive use of Equation 3.4 even in situations where its applicability is questionable.

In any event the chemical significance of the co-ordinate $x$ and the mass $m$ need careful examination, as may be illustrated by considering the reaction

$$XY + Z \rightarrow X + YZ$$

where X, Y and Z represent atoms or groups of atoms. If we consider only linear configurations, at any stage of the reaction process the system is defined by two distances, $R_{XY}$ and $R_{YZ}$, shown in the diagram,

$$X \underset{R_{XY}}{\text{———}} Y \underset{R_{YZ}}{\text{———}} Z$$

The simplest case is when X and Z are at a fixed distance apart, as in a rigid solid lattice. There is then only one distance which can vary independently (since $R_{XY} + R_{YZ} = R_{XZ} = \text{const.}$) and either $R_{XY}$ or $R_{YZ}$ can be taken as $x$. In either case the mass $m$ in the barrier problem is just $m_Y$. The same situation arises in an intramolecular process in which X and Z form part of a rigid molecular framework.

The situation is different if X, Y and Z are free to move in space. Again considering only linear configurations, $R_{XY}$ and $R_{YZ}$ are no longer independent, and the potential energy is now represented by a surface in three-dimensional space in which the co-ordinates are $V$, $R_{XY}$ and $R_{YZ}$. This is frequently depicted in the form of a map in which contours of constant $V$ are drawn in the $R_{XY} - R_{YZ}$ plane. Fig. 3.2 shows such a map for the particular

---

* Here and subsequently the term *semi-classical* is used to describe a treatment in which quantum corrections are applied to vibrations and (if necessary) to rotations, but not to translations; i.e. no account is taken of the tunnel effect.

# The application of tunnel corrections

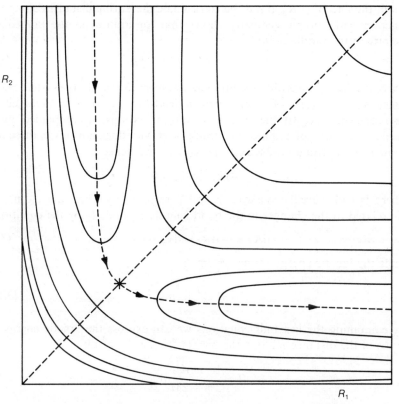

Fig. 3.2 Energy contour, map for reaction $X_1Y + X_2 \rightarrow X_1 + YX_2$

case in which X and Z are identical ($X_1$ and $X_2$) and the distances $X_1Y$ and $YX_2$ are termed $R_1$ and $R_2$.*

The course of the reaction is represented by the broken line, which is the path of lowest energy leading from the valley representing $X_1Y + X_2$ to that representing $X_1 + YX_2$, passing over the saddle point marked with an asterisk in the figure. In applying a one-dimensional tunnel correction it is usual to 'straighten out' this path and to plot the potential energy against a single co-ordinate representing motion along it, again a procedure of limited validity (cf. Chapter 7).

Since we are supposing $X_1$ and $X_2$ to be identical the diagram is symmetrical

---

* If the kinematics of the reaction are to be represented correctly by the motion of a particle on the energy surface it is necessary to use skewed co-ordinates and to plot distances on a different scale along the two axes. The angle between the co-ordinates and the scale factors are determined by the ratios of the masses of X, Y and Z [83]. However, these refinements are not needed for our present purposes.

about the line $R_1 = R_2$. If $\rho$ is the co-ordinate measuring distances along the reaction path, simple geometry shows that for motion in the immediate vicinity of the saddle point

$$\dot{\rho}^2 = \dot{R}_1^2 + \dot{R}_2^2 = 2\dot{R}_1^2 = 2\dot{R}_2^2 \tag{3.7}$$

where the dot signifies differentiation with respect to time. In the system being considered the type of motion which leads to reaction involves equal displacements $dx_X$ of the two X atoms (or groups) accompanied by a displacement $dx_Y$ of Y. Since the centre of gravity does not move, these are related by the equation $2M dx_X + m dx_Y = 0$, and hence

$$2M\dot{x}_X + m\dot{x}_Y = 0 \tag{3.8}$$

where $M$ and $m$ are the masses of X and Y, respectively. These displacements are related to the changes in the bond lengths $R_1$ and $R_2$ by the equation

$$dR_1 = -dR_2 = dx_Y - dx_X \tag{3.9}$$

while the kinetic energy of the system is

$$T = M\dot{x}_X^2 + \frac{1}{2}m\dot{x}_Y^2. \tag{3.10}$$

By combining the last four equations we can express the kinetic energy in terms of the distance $\rho$ along the barrier, giving the result

$$T = \frac{1}{2}\frac{mM}{m + 2M}\dot{\rho}^2. \tag{3.11}$$

This shows that if we are representing the kinetic energy of the system in terms of the motion of a single particle with velocity $\dot{\rho}$, the particle must be assigned an effective mass of $m_e = mM/(m + 2M)$, and it is therefore this mass which must be used in the tunnelling problem when solving the Schrödinger equation in terms of the co-ordinate $\rho$. If $M \gg m$, $m_e = \frac{1}{2}m$: this is approximately the situation when a hydrogen atom or proton moves between two much heavier atoms or groups. On the other hand, in the symmetrical reaction $H + H_2 \rightarrow H_2 + H$ we have $M = m$, and hence $m_e = \frac{1}{3}m$.

The same treatment can easily be generalized to cover three-particle systems less symmetrical than that considered above [84]. For the linear system X——Y——Z any type of motion can be characterized by a parameter $c$ defined by

$$\frac{dR_{YZ}}{dR_{XY}} = c \tag{3.12}$$

and equal to the slope of the trajectory in the plane $R_{XY} - R_{YZ}$. The effective mass for motion along this trajectory is then given by

$$m_e = \frac{m_X m_Y (1+c)^2 + m_Y(m_X + c^2 m_Z)}{(m_X + m_Y + m_Z)(1+c^2)}. \tag{3.13}$$

In the special case considered already, with identical X and Z, $c = -1$ at the top of the barrier from considerations of symmetry. The same value of $c$ applies if both $m_X$ and $m_Z$ are much larger than $m_Y$, and in that event we shall always have $m_e = \frac{1}{2} m_Y$. However, in general the location of the saddle point and the corresponding value of $c$ depend on the forces acting in the system. It should be noted that the effective mass remains constant only as long as the motion is represented by a straight line in the $R_{XY}$–$R_{YZ}$ diagram, i.e. over a small portion of the path in the immediate vicinity of the saddle point. The assumption of a constant effective mass over a larger section of the barrier is not strictly justifiable, though it is often made.

For a few very simple systems, notably that consisting of three hydrogen atoms (which may include any of the three isotopes of hydrogen), it is possible to use quantum theory to calculate the form of the energy surface from first principles, and hence the energy contour along the reaction path. More frequently the form of the barrier is evaluated in terms of a simple model, or by some more or less empirical treatment related, sometimes tenuously, to quantum theory. Some examples of these procedures are given in Chapter 5 in connection with individual gas reactions. For reactions in solution the commonest procedure is to assume the general form of the barrier (e.g. parabolic, Gaussian, Eckart) and to adjust its parameters so as to fit the experimental results as well as possible.

No mention has been made so far of zero-point energies in the reactant XY or the transition state X——Y——Z. XY must possess at least its vibrational zero-point energy, and it seems reasonable to raise the initial part of the potential energy profile by this amount [85], though this has been queried by some authors [86]. Detailed dynamical calculations for the reaction $H + H_2$ [87] suggest that only a part of the zero-point vibrational energy is available for crossing the barrier, the remainder being 'wasted' [88] in the symmetrical vibration of the transition state $\overleftarrow{X}$——Y——$\overrightarrow{Z}$: if non-linear configurations are included some of the energy is transferred to bending vibrations of the activated complex. A similar ambiguity exists as to whether the zero-point energies of the bending and symmetrical stretching vibrations of X——Y——Z should be added to the energy at the top of the barrier. In any treatment based on transition state theory, which assumes an equilibrium distribution of energy between degrees of freedom, it seems inevitable that the energy profile used for calculating tunnel corrections should include all zero-point energies of the initial and transition states, and also of intermediate configurations. In view of our ignorance of the exact form of energy surfaces the question is usually unimportant except when considering isotope effects, and it will be referred to again in that connection (Section 4.3).

Equation 3.6 shows that when the tunnel correction is small it depends only on the value of the barrier frequency $v_{\ddagger}$, as defined in Equation 2.40, and we shall see later that for a parabolic barrier the same is true even when the correction is moderately large. In the neighbourhood of the saddle point the energy surface for the system X——Y——Z can be represented by the quadratic expression

$$V = V_0 + \frac{1}{2}f_1(\Delta R_{XY})^2 + \frac{1}{2}f_2(\Delta R_{YZ})^2 + f_{12}\Delta R_{XY}\Delta R_{YZ} \qquad (3.14)$$

where $\Delta R$ signifies a departure from the value at the saddle point, i.e. at the top of the barrier. The force constants must satisfy the condition $f_{12}^2 > f_1 f_2$, since otherwise the system X——Y——Z would represent a stable species rather than a transition state. These force constants often emerge from a theoretical treatment, and in that event the barrier frequency $v_{\ddagger}$ can be derived without any separate consideration of the reaction path or the effective mass. The general expression for this frequency is

$$\left. \begin{array}{l} 4\pi^2 (iv_{\ddagger})^2 = \dfrac{1}{2}[B - (B^2 + 4C)^{1/2}] \\[2mm] B = \dfrac{f_1}{m_X} + \dfrac{f_2}{m_Z} + \dfrac{f_1 + f_2 - 2f_{12}}{m_Y} \\[2mm] C = \dfrac{(f_{12}^2 - f_1 f_2)(m_X + m_Y + m_Z)}{m_X m_Y m_Z} \end{array} \right\} \qquad (3.15)$$

which is obtained by applying the standard methods of vibrational analysis to the linear species X——Y——Z, when one of the frequencies turns out to have the imaginary value $iv_{\ddagger}$. Equation 3.15 can be much simplified in special cases. Thus if X and Z are identical, so that $m_X = m_Z = M$, and $f_1 = f_2 = f$, it becomes

$$4\pi^2 (iv_{\ddagger})^2 = -(f_{12} - f)\left(\frac{1}{M} + \frac{2}{m_Y}\right). \qquad (3.16)$$

All the above considerations apply to a transfer reaction of the type

$$XY + Z \rightarrow X + YZ.$$

This is in fact the most useful model for most chemical applications of the tunnel effect, since it includes the transfer of protons, hydrogen atoms or hydride ions between two species. Analogous methods can easily be developed for other types of reaction such as a dissociation

$$XYZ \rightarrow XY + Z$$

or a unimolecular isomerization

$$XYZ \rightarrow XZY.$$

## The application of tunnel corrections

The quantity $Q_t$ introduced in Equation 3.4 represents a tunnel correction to the classical expression for the flux of a Boltzmann distribution of particles across an energy barrier. In chemical applications it is necessary to select a classical or semi-classical expression for the reaction velocity which, when multiplied by $Q_t$, will give the actual velocity. The commonest procedure has been to use the formulation of *transition state theory*, though many of the general results are independent of this particular choice. For the present we shall accept this procedure, deferring until Chapter 7 an examination of its validity.

For a bimolecular reaction $A + B \rightarrow$ products, the observed rate can be expressed as $kc_A c_B$, where $k$ is the second-order velocity constant and $c_A$ and $c_B$ the volume concentrations of the reactants. The standard transition state expression for the velocity constant is then

$$(k)_s = \kappa \frac{kT}{h} \frac{F_\ddagger}{F_A F_B} \exp(-\varepsilon_0/kT) \qquad (3.17)$$

where $(k)_s$ denotes the semi-classical value, i.e. the value calculated without allowing for the tunnel effect. In Equation 3.17 $F_A$ and $F_B$ are the partition functions per unit volume for the reactants, and $F_\ddagger$ is the incomplete partition function per unit volume for the transition state, the degree of freedom corresponding to reaction (motion along the reaction co-ordinate) being omitted. The transition state resembles a normal molecule in all its degrees of freedom except motion along the reaction co-ordinate, which resembles a translation rather than a vibration, and any formal definition of the transition state must include an arbitrary small distance $\delta$ along this co-ordinate. This distance occurs in the translational partition function for motion along the reaction co-ordinate (not included in $F_\ddagger$), but cancels out when the number of systems in the transition state is divided by the average time taken to pass through it, leaving the universal factor $\mathbf{k}T/h$. The energy $\varepsilon_0$ is the energy required to pass from the lowest rotational and vibrational levels of A and B to the lowest level of the transition state. These lowest levels include the zero-point energies of any vibrational degrees of freedom, so that $\varepsilon_0$ will not coincide with the energy needed to pass from the reactants to the saddle point in the energy surface as usually represented, since this representation ignores the existence of zero-point energy.

The factor $\kappa$ in Equation 3.17 is a transmission coefficient, less than unity, which allows for the fact that some systems of sufficient energy may fail to cross the saddle point, but instead revert to the original reactants. This factor arises from the application of classical mechanics to the motion of a particle on the energy surface, and has nothing to do with the tunnel correction or the quantum-mechanical penetration of barriers by particles. Even when the form of the energy surface is known it may be difficult to estimate the value of $\kappa$, but there is reason to believe that it is close to unity for reactions in which there are no transitions from one quantum state to another.* It has therefore become usual to put $\kappa = 1$, and hence to omit it from Equation 3.17. The final expression for the observed velocity constant then becomes

$$k = Q_t(k)_s = Q_t \frac{kT}{h} \frac{F_\ddagger}{F_A F_B} \exp(-\varepsilon_0/kT). \qquad (3.18)$$

Analogous expressions for unimolecular and termolecular reactions are obtained by deleting or adding one partition function in the product $F_A F_B$.

## 3.3 The tunnel correction for a parabolic barrier

All realistic types of barrier can be approximated by a parabola in the neighbourhood of their maximum, just as all real oscillators can be treated as simple harmonic for small displacements. Moreover, the parabolic barrier yields a simple exact expression for the permeability (Equation 2.41), and since this expression contains the energy $W$ to the first power in the exponent, it is possible to evaluate the integral in Equation 3.4 explicitly. This model is therefore particularly suitable for investigating the general consequences of tunnel corrections, and it will be treated in some detail in this section.

Equation 2.41 applies to an infinite parabolic barrier, but in order to set a lower limit on the range of energies concerned the barrier must be truncated, as in Fig. 2.6. The barrier can then be assigned a definite width $a$ and height $E$, and the relations of these quantities to the barrier frequency $v_\ddagger$ and the force constant $A$ in the equation $V(x) = E - \tfrac{1}{2}Ax^2$ are given by Equations 2.40, 2.42 and 2.43. It is not strictly correct to use Equation 2.41 for the truncated barrier, but it is physically reasonable to suppose that the error will be small unless the main contribution to reaction comes from energies near the base of the parabola, i.e. unless the tunnel correction is very large: this is borne out by the results of numerical calculation [52, 53]. The evaluation of Equation 3.4 for a parabolic barrier was first given by Bell [81], though with some reservations about the range of validity of the result. In fact these reservations are unnecessary, as shown in Appendix C, which also contains a neater derivation than that originally given.

The expression obtained for $Q_t$ is

$$Q_t = \frac{\pi\alpha/\beta}{\sin(\pi\alpha/\beta)} - \alpha e^\alpha \left( \frac{e^{-\beta}}{\beta - \alpha} - \frac{e^{-2\beta}}{2\beta - \alpha} + \frac{e^{-\beta}}{3\beta - \alpha} - \cdots \right) \qquad (3.19)$$

---

* At one time values of $\kappa$ less than unity were frequently quoted even for the simplest reaction H + $H_2$ and its isotopic variants. This is because early attempts to calculate the energy surface from approximate versions of the quantum theory predicted the existence of a shallow basin, i.e. an unstable triatomic complex, in the vicinity of the saddle point: this complex would have comparable probabilities of going on to form products or reverting to reactants. However, more refined quantum-mechanical treatments do not predict any such basin, which is now believed to have been an artefact of the approximations used in the early calculations.

where
$$\alpha = E/kT, \quad \beta = 2\pi/h\nu_{\ddagger}. \tag{3.20}$$

An alternative way of writing Equation 3.19 is

$$Q_t = \frac{\tfrac{1}{2}u_{\ddagger}}{\sin \tfrac{1}{2}u_{\ddagger}} - u_{\ddagger}y^{-u_{\ddagger}/2\pi}\left(\frac{y}{2\pi - u_{\ddagger}} - \frac{y^2}{4\pi - u_{\ddagger}} + \frac{y^3}{6\pi - u_{\ddagger}} - \cdots\right) \tag{3.21}$$

in which

$$u_{\ddagger} = h\nu_{\ddagger}/kT$$
$$y = \exp(-2\pi E/u_{\ddagger}kT) = \exp(-2\pi\alpha/u_{\ddagger}) = \exp(-\beta). \tag{3.22}$$

If $\alpha = m\beta$ ($u_{\ddagger} = 2m\pi$), where $m$ is an integer, both the first term and one term in the series in these expressions become infinite. It is shown in Appendix C that the sum of these two terms remains finite and has the value $(-1)^{m+1}\alpha$. In making the calculations it may be more convenient in practice to avoid values of $\alpha/\beta$ which are close to an integer and to interpolate the values of $Q_t$ in these regions.

In problems of chemical interest the values of $\alpha$ and $\beta$ are almost always > 5, and usually > 10. The quantities $e^{-\alpha}$ and $e^{-\beta}$ are thus normally much less than unity, and $e^{\alpha-\beta}$ therefore usually either $\gg 1$ or $\ll 1$. The second possibility is more commonly met with, and implies that only the first term of Equation 3.19 or Equation 3.21 need be retained. The resulting expression (Equation 3.23) is the one which has most commonly been used for evaluating small or moderate tunnel corrections in chemical kinetics. It implies that $\alpha < \beta$, or $u_{\ddagger} < 2\pi$.

$$Q_t = \frac{\pi\alpha/\beta}{\sin(\pi\alpha/\beta)} = \frac{\tfrac{1}{2}u_{\ddagger}}{\sin\tfrac{1}{2}u_{\ddagger}}. \tag{3.23}$$

Equation 3.23 demands only a knowledge of the barrier frequency $\nu_{\ddagger}$, i.e. of the curvature of the barrier at the top and the effective mass of the particle, while the subsequent terms in Equations 3.19 or 3.21 require in addition the barrier height $E$, or alternatively the barrier width, which can be combined with the curvature to give $E$. Expansion of Equation 3.23 in powers of $u_{\ddagger}$ gives

$$Q_t = 1 + \frac{u_{\ddagger}^2}{24} + \frac{7u_{\ddagger}^4}{5760} + \cdots \tag{3.24}$$

$$\ln Q_t = \frac{u_{\ddagger}^2}{24} + \frac{u_{\ddagger}^4}{2880} + \cdots. \tag{3.25}$$

The first term in these expansions agrees with Wigner's expression (Equation 3.6) for small tunnel corrections.

Equation 3.23 shows a remarkable analogy to the quantum correction to the quasi-classical expression for the partition function of a harmonic oscillator of frequency $v$. The quasi-classical value is given by

$$F' = \frac{1}{h}\int_{-\infty}^{+\infty}\int_{-\infty}^{+\infty} \exp[W(p_x, x)/kT]dp_x dx$$

$$= \frac{1}{h}\int_{-\infty}^{+\infty}\int_{-\infty}^{+\infty} \exp\left(\frac{p_x^2}{2mkT} + \frac{2\pi^2 m^2 v^2 x^2}{kT}\right)dp_x dx \qquad (3.26)$$

$$= \frac{kT}{hv} = \frac{1}{u}$$

where $u = hv/kT$. The exact quantum-mechanical value (reckoned from the same energy zero) is

$$F = e^{-u/2}(1 + e^{-u} + e^{-2u} + \cdots)$$

$$= \frac{e^{-u/2}}{1 - e^{-u}} = \frac{1}{2\sinh\frac{1}{2}u} \qquad (3.27)$$

so that the correction factor introduced by vibrational quantization is

$$\frac{F}{F'} = \frac{\frac{1}{2}u}{\sinh\frac{1}{2}u} \qquad (3.28)$$

Since $\sinh ix = i\sin x$, Equation 3.28 becomes identical with Equation 3.23 if the real frequency $v$ is replaced by the imaginary frequency $iv_\ddagger$. Similarly, the expansion of $F/F'$ in powers of $u$ is identical with Equation 3.24, except that the terms are alternately positive and negative instead of being all positive. This correspondence has been described as 'an amusing coincidence' [89], but may conceal some fundamental connection which is not yet fully understood. Several authors [49, 90–92] have suggested general procedures for deriving the tunnel correction for a one-dimensional barrier from the energy spectrum for the motion of a particle in the potential well obtained by inverting the barrier. Their recipes appear to rest on intuition rather than on strict logic, though they yield the correct result in particular cases where a test is possible.

In any event the parallelism between Equations 3.23 and 3.28 reinforces the point that it is quite illogical to apply quantum corrections to the real vibrations of the transition state but to omit the tunnel correction. Both corrections depend fundamentally on the uncertainty principle, and if $v$ and $v_\ddagger$ have similar values the tunnel correction will be at least as important as the correction to the vibrational partition function, though in the opposite direction.

Equation 3.19 can also be simplified if $e^{\alpha - \beta} \gg 1$. This situation is not commonly encountered in chemical kinetics, but will always arise if the temperature is sufficiently low. The second part of Equation 3.19 then becomes dominant, and provided that $e^\beta \gg 1$ only the first term of the series need be retained, giving the result

$$Q_t = \frac{\alpha \exp(\alpha - \beta)}{\alpha - \beta}. \qquad (3.29)$$

# The application of tunnel corrections

This expression is valid when the tunnel correction is very large, but does not apply to small corrections, and does not reduce to Wigner's result (Equation 3.6) when expanded in powers of $h/kT$. The same expression was deduced many years ago [93] on the assumption that $G(W)$ is unity for $W > E$, and is given by the commonly used BWK approximation (Equation 2.90) for $W < E$. These approximations do not cause an appreciable error when most of the reaction involves systems with energies well below the top of the barrier, but are invalid if the departures from classical behaviour are small.

If neither of the approximations $e^{\alpha - \beta} \ll 1$ or $e^{\alpha - \beta} \gg 1$ is permissible, or if $\alpha/\beta$ is close to an integer, both parts of Equation 3.19 must be included in calculating the tunnel correction, though it is often possible to neglect the terms in $e^{-2\beta}$ and higher powers.

## 3.4 Tunnel corrections for other types of barrier

The explicit evaluation of Equation 3.4 has not proved possible for any of the other types of barrier considered in Chapter 2, and this applies both to the exact expressions for $G(W)$ and to those obtained by using the BWK approximation.* The value of $Q_t$ can of course be obtained by numerical or graphical integration for any type of barrier for given values of the barrier parameters, the particle mass and the temperature. This is most readily achieved when an explicit expression is available for $G(W)$, but can also be done when $G(W)$ itself has to be obtained by numerical methods (cf. Section 2.4.6), for example for the Gaussian barrier (Equation 2.68).

Because of their realistic form and the availability of exact expressions for $G(W)$ attention has been concentrated on the computation of $Q_t$ for Eckart barriers (Section 2.4.1). Tables of $Q_t$ have been given by several authors [94, 95] and an analytical approximation, valid under certain conditions, has also been derived [96]. The general results of such calculations, in terms of the dependence of $Q_t$ upon temperature, barrier parameters and particle mass, resemble closely those derived by assuming the barrier to be parabolic. Since the latter can be expressed in equations which are easy to handle and appreciate, much of the subsequent treatment in this book will be based on parabolic barriers, though it should be stressed that the extra labour involved in employing more realistic types of barrier will be justified if information is available about the true shape of the barrier, or if large tunnel corrections are involved. Examples of such calculations for particular systems will be quoted in Chapter 5.

## 3.5 Activation energies and pre-exponential factors

The effect of tunnelling in chemical kinetics appears more clearly when the

---

* An explicit expression for $Q_t$ could in fact be obtained for the double anharmonic barrier (Section 2.4.3), but has not been applied to any physical or chemical problems.

temperature dependence of reaction velocity is examined than when only velocities at a single temperature are considered. It is usual to analyse this dependence in terms of the *Arrhenius equation* (Equation 3.30), which is found empirically to hold over at least a limited range of temperature.

$$k = (A)_A \exp[-(E)_A/kT]. \tag{3.30}$$

The subscript A in this equation means that the quantities in question have been derived on the basis of the Arrhenius law: they will not in general be identical with the corresponding quantities in more elaborate theoretical or empirical relations. The operational definition of $(E)_A$ is thus

$$(E)_A = kT^2 d\ln k/dT = -k d\ln k/d(1/T) \tag{3.31}$$

where $k$ is the observed velocity constant, and the Arrhenius pre-exponential factor is obtained by substituting $(E)_A$ in Equation 3.30, giving

$$\ln(A)_A = \ln k + (E)_A/kT. \tag{3.32}$$

Since $k = Q_t(k)_s$ the above equations become

$$(E)_A - (E)_s = kT^2 d\ln Q_t/dT \tag{3.33}$$

$$\ln\frac{(A)_A}{(A)_s} = \ln Q_t + T\frac{d\ln Q_t}{dT} = \frac{d}{dT}(T\ln Q_t) \tag{3.34}$$

in which the semi-classical activation energy and pre-exponential factor are defined by

$$(E)_s = kT^2 \frac{d\ln(k)_s}{dT}, \quad \ln(A)_s = \ln(k)_s + \frac{(E)_s}{kT}. \tag{3.35}$$

Equations 3.33 and 3.34 are of course valid independent of any particular barrier shape or expression for $Q_t$. Since $d\ln Q_t/dT$ is always negative, Equation 3.33 predicts that the effect of tunnelling will be to make the Arrhenius activation energy smaller than the semi-classical value. This lowering of the activation energy has a simple physical interpretation. It can be shown by very general statistical considerations [97] that the Arrhenius activation energy, as defined by Equation 3.31, is equal to the difference between the average energy of the reacting systems and the average energy of all the systems. The effect of tunnelling is to lower the average energy of the reacting systems (cf. Fig. 3.1) and hence to decrease $(E)_A$.

The direction in which tunnelling affects the pre-exponential factor is not so immediately obvious, since on the right-hand side of Equation 3.34 $\ln Q_t$ is positive but $T d\ln Q_t/dT$ is negative. However, the general treatment given by Wigner [38] shows that if $\ln Q_t$ is expanded in powers of $h/kT$ the lowest term is positive and proportional to $T^{-2}$ (Equation 3.6): hence for small tunnel corrections we shall always have $(A)_A < (A)_s$. It seems safe to conclude that this

# The application of tunnel corrections

will still be true when the correction is no longer small, and this is borne out by the results for the parabolic barrier which follow shortly.

The finding that $(A)_A < (A)_s$ is the opposite of what might be expected intuitively. The way in which it arises is shown schematically in Fig. 3.3, in which both $\ln k$ and $\ln k_s$ are plotted against $1/T$. The procedure for obtaining $\ln(A)_A$ implied by Equations 3.31 and 3.32 is to draw the best straight line through the experimental values of $\ln k$ over a limited temperature range: $\ln(A)_A$ is then equal to the intercept on the $\ln k$ axis which this straight line makes at $1/T = 0$. Since the amount by which $\ln k$ exceeds $\ln(k)_s$ always increases with decreasing temperature it is clear from the figure that this intercept will be smaller than the semi-classical value $\ln(A)_s$.

We shall now consider the expressions obtained for a parabolic barrier, taking first the case of slight or moderate tunnelling, which occurs most frequently in chemical applications. Substitution of Equation 3.23 in Equations 3.33 and 3.34 leads to

$$(E)_A - (E)_s = -\mathbf{k}T(1 - \tfrac{1}{2}u_{\ddagger}\cot\tfrac{1}{2}u_{\ddagger}) \tag{3.36}$$

$$\ln\frac{(A)_A}{(A)_s} = \ln\frac{\tfrac{1}{2}u_{\ddagger}}{\sin\tfrac{1}{2}u_{\ddagger}} + \tfrac{1}{2}u_{\ddagger}\cot\tfrac{1}{2}u_{\ddagger} - 1 \tag{3.37}$$

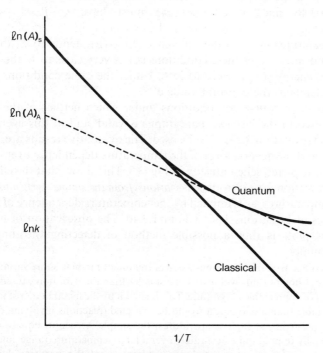

Fig. 3.3  Effect of tunnelling upon Arrhenius parameters

which are valid provided that $u_{\ddagger} < 2\pi$ and $e^{\alpha-\beta} \ll 1$. These expressions agree with the qualitative statements made in the last two paragraphs, as may be seen by using the series expansions (Equations 3.24 and 3.25), which give

$$(E)_A - (E)_s = -\mathbf{k}T(\tfrac{1}{12}u_{\ddagger}^2 + \tfrac{1}{720}u_{\ddagger}^4 + \cdots) \tag{3.38}$$

$$\ln\frac{(A)_A}{(A)_s} = -(\tfrac{1}{24}u_{\ddagger}^2 + \tfrac{1}{960}u_{\ddagger}^4 + \cdots). \tag{3.39}$$

Each of the last four expressions goes to zero as $T \to \infty$; i.e. the effect of tunnelling disappears at a sufficiently high temperature.

At the other extreme, corresponding to large tunnel corrections, we can use Equation 3.29, valid when both $e^{\beta}$ and $e^{\alpha-\beta}$ are much greater than unity. Insertion in Equations 3.33 and 3.34 gives

$$(E)_A - (E)_s = -E\left[1 - \frac{\beta}{\alpha(\alpha-\beta)}\right] \tag{3.40}$$

$$\ln\frac{(A)_A}{(A)_s} = -\left[\ln\left(1 - \frac{\beta}{\alpha}\right) + \beta\left(1 - \frac{1}{\alpha-\beta}\right)\right]. \tag{3.41}$$

Under the appropriate conditions each of these expressions is negative, again agreeing with the qualitative statements already made. This is obvious for Equation 3.40 since if $e^{\alpha-\beta} \gg 1$ we must have $\alpha - \beta > 1$, while for Equation 3.41 it can be demonstrated by expanding in powers of $\beta/\alpha$. At very low temperatures ($\alpha \to \infty$) the difference $(E)_s - (E)_A$ tends to $E$, the barrier height, and since under these conditions $(E)_s$ is very close to $E$, the observed activation energy $(E)_A$ is close to zero. Under the same conditions the ratio $(A)_A/(A)_s$ tends to the constant value $e^{-\beta}$.

There will of course be conditions under which neither of the approximations used in the last two paragraphs is valid, and in this event the full expression (Equation 3.21) must be used. The qualitative results are, however, the same, and the general form of the temperature dependence over the whole range is as depicted schematically in Fig. 3.3. This shows that the effect of the tunnel correction is to introduce *deviations from the simple Arrhenius equation* at low temperatures, as is implied by the temperature dependence of the right-hand sides of Equations 3.36, 3.38 and 3.40. The observation of non-linear Arrhenius plots is thus a possible method of detecting the tunnel effect experimentally.

It is of course true that the semi-classical treatment predicts some variation of $(E)_s$ with temperature, but this will usually be smaller than the variation caused by tunnel corrections of appreciable magnitude. This is shown by the fact that in instances where no appreciable tunnel correction would be expected (reactions involving only heavy atoms, or occurring at high temperatures) the simple Arrhenius equation is found experimentally to hold quite accurately over a large temperature range, and the same conclusion is reached by a detailed consideration of the partition functions in

# The application of tunnel corrections

Equation 3.17. The point can again be illustrated by considering the equations for parabolic potential energy curves. It was shown in Section 3.3 (Equations 3.26–3.28) that there is a close analogy between the tunnel correction for a parabolic barrier and the quantum correction to the partition function of a harmonic oscillator. A similar analogy exists between the tunnel correction to the activation energy and the quantum correction to the mean energy of a harmonic oscillator in the initial or transition state: in fact, these two corrections can be interconverted by replacing $iv_\ddagger$ in Equation 3.36 or Equation 3.38 by $v$, the real vibration frequency. If $v$ and $v_\ddagger$ have similar numerical values, then provided Equation 3.38 can be approximated by its first term the two corrections will be of similar magnitude but in opposite directions. This means that it will be difficult to attribute small deviations from the Arrhenius equation to tunnelling, since the overall effect of the real frequencies represents a difference between terms relating to the transition state and to the initial state respectively, and hence may be in either direction. However, if the first term of Equation 3.38 is not sufficient, the tunnel correction tends to predominate, since all the terms in Equation 3.38 have the same sign, while in the corresponding expression for a real oscillator they are alternately positive and negative. In fact, for a real oscillator the limiting low-temperature correction to the average energy is $\frac{1}{2}hv$, the zero-point energy, while at low temperatures the tunnel correction gives rise to the much larger contribution given by Equation 3.40.

Detailed calculations on the effect of tunnelling on the curvature of Arrhenius plots have been carried out by Stern and Weston [98] on the basis of one-dimensional symmetrical Eckart barriers and a wide range of barrier parameters and temperatures. Their general conclusion is that there is no simple relationship between this curvature and the extent of tunnelling, though curvature is favoured by high barriers and values of $u_\ddagger$ around 8.

There is also a close parallelism between the effect of tunnelling on $\ln A$ (Equations 3.37 and 3.39) and the quantum correction to the entropy of a harmonic oscillator. The distinction between the two effects is again difficult when the tunnel correction is small, but becomes more marked for moderate and large degrees of tunnelling. The effect of tunnelling on pre-exponential factors is mainly of importance in connection with isotope effects, and will be discussed in Chapter 4.

## 3.6 General criteria for tunnelling

For many purposes it is useful to classify tunnelling problems according to the magnitude of the tunnel correction. Such subdivisions are of course arbitrary, but the following three categories are often useful.

*Negligible tunnelling*, $1 < Q_t < 1.1$, i.e. tunnel corrections amounting to less than 10%. Such small corrections will rarely be detectable experimentally and can therefore usually be omitted from a theoretical treatment.

*Small to moderate tunnelling*, $1.1 < Q_t < 4$. This is the range most commonly met with in chemical reactions at ordinary temperatures. Tunnelling must be allowed for in a quantitative treatment, but will not give rise to any qualitatively striking phenomena.

*Large tunnelling*, $Q_t > 4$. This range has no upper limit. It will rarely be encountered in chemical reactions at ordinary temperatures, but will always appear if the temperature is lowered sufficiently. The value of $Q_t$ can then reach many powers of ten, and the behaviour of the system can differ qualitatively from that usually encountered: it then becomes reasonable to say that the process is occurring by a 'tunnelling mechanism'.

### 3.6.1 Tunnelling criteria for parabolic barriers

If the barrier is parabolic, the conditions for the three categories of tunnelling in terms of the value of $u_{\ddagger}$ ($\equiv hv_{\ddagger}/kT \equiv 2\pi\alpha/\beta$) follow directly from Equations 3.19–3.29.

*Negligible tunnelling* occurs when $u_{\ddagger} < 1.5$ (in round figures). On the rare occasions when it is necessary to calculate the small corrections involved $Q_t$ is given with ample accuracy by the first term of Equation 3.24 or Equation 3.25, which then become identical with the general expression Equation 3.6 derived by Wigner.

*Small to moderate tunnelling* corresponds to the range $1.5 < u_{\ddagger} < 5$ (again in round figures). Throughout this range, if $\alpha$ and $\beta$ have the values characteristic of chemical problems, $Q_t$ is given with fair accuracy by the simple expression (Equation 3.23), and is therefore a function of $u_{\ddagger}$ alone. A useful 'marker' in the middle of this range is given by the special case $u_{\ddagger} = \pi (\alpha = 2\beta)$, when $Q_t = \frac{1}{2}\pi = 1.57$.*

*Large tunnelling* occurs when $u_{\ddagger} > 5$. In this range it is necessary to use both parts of Equation 3.19 or Equation 3.21, so that $Q_t$ is no longer a function of $u_{\ddagger}$ alone, but requires any two of the quantities $u_{\ddagger}$, $\alpha$ and $\beta$. This means that for a particle of a given mass at a given temperature the tunnel correction depends not only on the curvature of the barrier, but also on its height (or width): in physical terms this is because an appreciable proportion of the reacting systems have energies corresponding to the lower part of the barrier. The point is illustrated by Table 3.1, which shows the dependence of $Q_t$ upon the barrier height for the passage of a proton or hydrogen atom across barriers of varying curvature.†

Useful special cases applicable to this region can be derived from Equation

---

* As shown in Appendix C (Equation C.14) the exact solution for $u_{\ddagger} = \pi$ is $Q_t = \tan^{-1} e^{\alpha}$, but since in chemical problems the value of $\alpha$ is usually at least 5, we can use the asymptotic expansion $\tan^{-1} x = \frac{1}{2}\pi - x^{-1} + \frac{1}{3}x^{-3} - \ldots$, retaining only the first term.
† Values calculated by Professor E. F. Caldin, and reproduced by his kind permission.

Table 3.1 *Tunnel corrections for symmetrical parabolic barriers of varying height and curvature* Values refer to a proton or hydrogen atom and to 298 K.

| $u_\ddagger$ | $v(\text{cm}^{-1})$ | | | $E(\text{kJ mol}^{-1})$ | | | |
|---|---|---|---|---|---|---|---|
| | | 16.7 | 25.1 | 33.5 | 41.8 | 62.7 | 83.6 |
| 5   | 1036 | 3.49 | 3.90 | 4.06 | 4.13 | 4.17 | 4.18 |
| 6   | 1244 | 6.0  | 8.1  | 10.2 | 11.6 | 14.9 | 21   |
| 7   | 1451 | 9.3  | 17.4 | 29   | 45   | 121  | 302  |
| 7.5 | 1555 | 11.8 | 25   | 48   | 89   | 365  | 1457 |

3.19 or Equation C.14 (Appendix C). The results given in Equation 3.42 are valid provided that $e^{-\alpha}$, $e^{-\beta}$ and $e^{\beta-\alpha}$ are all much smaller than unity.

$$\left.\begin{array}{l} \alpha = \beta, \quad u_\ddagger = 2\pi, \quad Q_t = \alpha = \beta \\ \alpha = \dfrac{3\beta}{2}, \quad u_\ddagger = 3\pi, \quad Q_t = 3e^{\alpha/3} = 3e^{\beta/2} \\ \alpha = 2\beta, \quad u_\ddagger = 4\pi, \quad Q_t = 2e^{\alpha/2} = 2e^{\beta}. \end{array}\right\} \quad (3.42)$$

If we are considering a particular process the barrier parameters and the particle mass are fixed, and the degree of tunnelling depends on the temperature. The behaviour of the process as regards tunnelling can be conveniently represented by a *characteristic temperature* $T_c$. This approach has been used particularly by Christov [66, 99–101], who defines $T_c$ for a parabolic barrier by the equation

$$T_c = hv_\ddagger/\pi \mathbf{k} = hA^{1/2}/2\pi^2 m^{1/2}\mathbf{k} \quad (3.43)$$

where $A$ is the curvature of the barrier at its apex. The first term of Equation 3.19 or Equation 3.21 then becomes

$$\frac{\tfrac{1}{2}\pi(T_c/T)}{\sin[\tfrac{1}{2}\pi(T_c/T)]}$$

and the subsequent terms can also be expressed in terms of $T_c$.

At a temperature $T = T_c$, corresponding to $u_\ddagger = \pi$, the reacting systems are approximately equally divided between those with $W > V_0$ and those with $W < V_0$, as may be seen by integrating Equation 3.4 separately over the two ranges $V_0 < W < \infty$ and $0 < W < V_0$. The exact results are

$$Q'_t(W > V_0) = \frac{\pi}{4}$$

$$Q''_t(W < V_0) = \tan^{-1} e^\alpha - \frac{\pi}{4} = \frac{\pi}{4} - e^{-\alpha} + \frac{1}{3}e^{-3\alpha} - \ldots$$

(3.44)

where the last line employs the asymptotic expansion of $\tan^{-1} x$ for large $x$. Since $e^\alpha \gg 1$ for most systems of chemical interest it is normally a very good approximation to write $Q'_t = Q''_t$ when $T = T_c$.*

Christov shows further [100] that tunnelling will be negligible when $T > 2T_c$, moderate when $T_c > T > \frac{1}{2}T_c$, and large when $T < \frac{1}{2}T_c$. These divisions, which of course have no absolute quantitative significance, are approximately equivalent to the criteria in terms of $u_\ddagger$ given earlier in this section.

### 3.6.2. Model systems and the effect of transition state symmetry

It has been shown by Christov [66, 98, 99] that criteria similar to those derived for a parabolic barrier also apply to a more general type of barrier. However, the relevant parameters of the barrier (parabolic or otherwise) for an actual reaction can rarely be deduced independently of the observed kinetic behaviour, and the most usual procedure is to estimate these parameters by comparing theoretical predictions with experimental kinetic results. For a few very simple reactions in the gas phase, notably those involving the transfer of a hydrogen atom, it is possible to construct approximate energy surfaces by applying various approximate versions of the quantum theory, or semi-empirical procedures related thereto: some of these are referred to in Chapter 5. However, apart from absolute estimates of tunnel corrections, it is often of interest to know how they are related to other quantities, such as the endothermicity or exothermicity of the reaction, or other frequencies of the transition state. This can be explored by investigating models, which, although admittedly crude, are often amenable to exact treatment.

Most model calculations have been made for linear configurations of three-centre reactions of the type

$$XY + Z \rightarrow X + YZ$$

already discussed in Section 3.2. The problem is greatly simplified if it is assumed that X and Z remain at a fixed distance apart, while Y moves between them. This is approximately the case when the transfer of Y takes place in a solid lattice, or intramolecularly between two sites in a rigid molecular species. If exchange interactions can be neglected the force acting on Y is the resultant of the forces exerted by X and Z, which can be written as $\phi_X(r_{XY})$ and $\phi_Z(r_{ZY})$. In the transition state this resultant is zero, and since $r_{XY} + r_{ZY} = \text{const.} = 2L$, we can write

$$\phi_X[L(1+\gamma)] = \phi_Z[L(1-\gamma)] \tag{3.45}$$

---

* This does not imply that the tunnel correction is equal to two at this temperature, the correct value being $\frac{1}{2}\pi$: this is because the probability of crossing the barrier is less than unity even when $W > V_0$. It should be noted that somewhat different definitions of $T_c$ are given in different papers by Christov. That given in Equation 3.43 agrees with [100] and is the only one which leads to equal fluxes above and below the barrier when $T = T_c$. $T_c$ is also the temperature at which $(E)_A = (E)_s - kT$ and $(A)_A/(A)_s = \pi/2e = 0.59$, as may be seen from Equations 3.36 and 3.37.

# The application of tunnel corrections

where $\gamma$ characterizes the position of the transition state. It is then easy to show [102] that the effective force constants $F_3$ and $F_2$ for the motion of Y along and at right angles to the line of centres are given respectively by

$$F_3 = \phi'_X[L(1+\gamma)] + \phi'_Z[L(1-\gamma)] \tag{3.46}$$

$$F_2 = \frac{\phi_X[L(1+\gamma)]}{L(1+\gamma)} + \frac{\phi_Z[L(1-\gamma)]}{L(1-\gamma)} \tag{3.47}$$

in which $\phi' = d\phi/dr$. The force constant $F_3$ is negative, corresponding to the maximum in potential energy for motion along the reaction co-ordinate, while $F_2$ is positive and refers to the real bending frequency of the transition state. The laws of force in the vicinity of the transition state can be characterized by two positive numbers,

$$n_X = -d\ln\phi_X/dr, \quad n_Z = -d\ln\phi_Z/dr. \tag{3.48}$$

If $n_X = n_Z = n$ the last three equations give the simple result

$$F_3/F_2 = -n. \tag{3.49}$$

If $n_X$ and $n_Z$ are different the same result holds approximately for an average value $\bar{n} = \frac{1}{2}(n_X + n_Z)$, the exact expression being

$$F_3/F_2 = -\bar{n}[1 + \frac{1}{2}\gamma(n_X - n_Z)/\bar{n}]. \tag{3.50}$$

Since the end groups X and Z are assumed to remain stationary, the effective mass for motion along or perpendicular to the line of centres is just $m_Y$, and Equation 3.49 leads to the result

$$\frac{(iv_{\ddagger})^2}{v_2^2} = -n, \quad \frac{v_{\ddagger}}{v_2} = n^{1/2} \tag{3.51}$$

where $iv_{\ddagger}$ is the imaginary barrier frequency and $v_2$ the real bending frequency of the transition state. Since $n$ (or the mean value $\bar{n}$) is in practice a small number greater than two, Equation 3.51 predicts that $v_{\ddagger}$ of the same order of magnitude as $v_2$, their ratio having a minimum value of $2^{1/2}$. This finding is of some interest, since, although we cannot of course directly observe the bending frequency of the transition state, it can sometimes be estimated by analogy with the bending frequencies of stable species. For example, in proton-transfer reactions the bending frequencies of hydrogen-bonded species provide a reasonable model.

The above treatment is based upon a restricted model in which the end groups X and Z are fixed. The derivation becomes more complicated in the more realistic situation in which all three atoms or groups are free to move, since the frequencies are now given by expressions such as Equation 3.15, the effective mass by Equation 3.13, and it is also necessary to introduce repulsive forces between X and Z. However, it is found [102] that Equations 3.49 and 3.51 remain at least good approximations provided that $m_X$

and $m_Z$ are considerably greater than $m_Y$, a situation which obtains in the transfer of a proton, hydrogen atom or hydride ion between two heavier atoms or groups. Since it is just this last group of processes which is relevant when considering the tunnel effect, the simple treatment will often be adequate.

The assumption of central forces means that the above considerations apply primarily to electrostatic models, and these are particularly appropriate in considering proton-transfer reactions, for which (at least in the vicinity of the transition state) it is reasonable to picture the proton as moving in a force field determined by charges, permanent dipoles or induced dipoles on X and Z. If only charge–charge interactions are involved we have $n = 2$ in Equations 3.49 and 3.51, and hence $v_{\ddagger} = 2^{1/2} v_2$. It is of interest to compare the magnitudes of the tunnel correction and the quantum correction to the partition function for the bending vibration of the transition state. If both corrections are small we have seen in Section 3.3 (Equations 3.24–3.28) that they are given respectively by $1 + (1/24)(hv_{\ddagger}/kT)^2$ and $1 - (1/24)(hv_2/kT)^2$. The bending vibration is doubly degenerate, so that if terms in $(hv/kT)^4$ and higher powers can be neglected the two corrections will cancel one another. If the corrections are not small the tunnel correction will be the greater of the two, and this will be the case even for small corrections if $n > 2$, as for the interaction of the proton with permanent or induced dipoles. These conclusions must not be taken too seriously, but they do show once more that it is quite illogical to ignore the tunnel effect if quantum corrections are being applied to the vibrations of the transition state.

Any electrostatic model can be used to predict absolute values of $v_{\ddagger}$ and hence of the tunnel correction, and also of the bending frequency $v_2$. Although this procedure gives reasonable values, the result is sensitive to the dimensions of the model, and also to the choice of dielectric constant, which is to a large extent arbitrary. Greater interest attaches to the use of such models for investigating in a series of similar reactions the relationship between the tunnel correction and the *symmetry of the transition state*, or (which comes to the same thing) the endothermic or exothermic nature of the overall process. As a simple example we may consider the motion of a proton between charges of $-\beta e$ and $-(1 - \beta)e$ at a fixed distance $2L$ apart. If the transition state is distant $(1 + \gamma)L$ and $(1 - \gamma)L$ respectively, from these two charges the condition of equilibrium is

$$\beta(1 - \gamma)^2 = (1 - \beta)(1 + \gamma)^2 \tag{3.52}$$

and it is easily shown [102] that

$$(iv_{\ddagger})^2 = - e^2/2\pi^2 mDL^3(1 - \gamma^4) \tag{3.53}$$

where $m$ is the mass of the proton and $D$ the effective dielectric constant. Equation 3.53 shows that $v_{\ddagger}$ has a minimum value for a symmetrical transition state, for which $\gamma = 0$, so that if the tunnel correction is small enough to be

# The application of tunnel corrections

given with sufficient accuracy by Equation 3.23 it will have a *minimum* value when the transition state is symmetrical. However, in practice only small variations in $Q_t$ are likely to arise from this cause: thus if $\beta$ varies from 0.1 to 0.9 $\gamma$ goes from $-\frac{1}{2}$ to $+\frac{1}{2}$, and the consequent changes in $v_{\ddagger}$ are only about 3%. If, on the other hand, the tunnel correction is large enough to require the inclusion of the exponential terms in Equation 3.19 or Equation 3.21 much larger variations of $Q_t$ with transition state symmetry appear, and these are in the opposite direction, i.e. the tunnel correction has a *maximum* value when the transition state is symmetrical. This is because the exponential terms involve not only the barrier frequency $v_{\ddagger}$ but also the barrier height $E$, which, on any reasonable model, depends considerably on the position of the transition state. The tunnel correction must have the same value for both the forward and the reverse reaction, so that for an endothermic reaction the appropriate barrier height is equal to the height of the barrier for the reverse process. This is illustrated in Fig. 3.4, which shows that only that portion of the barrier which lies above both the initial and the final states is available for tunnelling. This portion is greatest when the reaction is thermoneutral, leading to a maximum tunnel correction for this configuration.

An electrostatic model involving only point charges does not yield any finite values for the barrier height or for the energy change in the reaction. (The latter is more directly related to experiment than is the position of the transition state, represented here by $\gamma$.) This failing can be remedied by siting the charges on spheres of fixed radius, supposed impenetrable to the proton [103, 105]. For typical values of the parameters the results again show that the tunnel correction passes through a clearly marked maximum when the transition state is symmetrical and the proton transfer thermoneutral. Similar conclusions have been reached on the basis of other electrostatic models and also by treating a more sophisticated model in which the proton moves in the field of two spherical distributions of negative charge [104, 105].

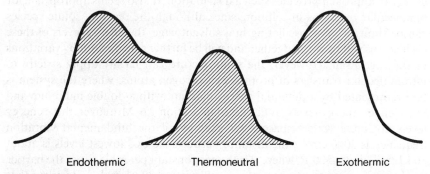

Fig. 3.4 Region available for tunnelling in endothermic, thermoneutral and exothermic reactions

The main application of the above findings is to the variation of hydrogen isotope effects in a series of similar proton-transfer reactions, which is discussed in Chapters 4 and 5. The tunnel correction should in principle affect the relation between rates and equilibrium constants for a series of similar acid–base reactions even when only one hydrogen isotope is involved. The simplest form of such relations, the Brönsted relation, consists of a linear dependence of ln $k$ upon ln $K$, where $k$ is the velocity constant and $K$ is equal or proportional to the equilibrium constant of the reaction: either the acid or the base may be acting as a catalyst. It is now clear that there will be departures from linearity if $K$ (and therefore $k$) is varied over a sufficiently wide range, and several recent reviews [106–108] have described the experimental evidence for such departures and their interpretation in terms of classical theory. As first pointed out many years ago [109, 110], the dependence of the tunnel correction on the equilibrium constant will cause further deviations from linearity which will be greatest when the equilibrium constant is unity, i.e. when the two systems reacting have the same acid–base strengths. Although there is one instance in which this effect has been cited to account for experimental findings [111], the evidence is not convincing. It seems unlikely that unequivocal evidence for tunnel effects can be obtained from Brönsted plots, since the predicted curvature is in the same sense as that expected on a classical basis, and it is not often possible to investigate a series which includes reactions which are both up-hill and down-hill in a thermodynamic sense. Convincing evidence could be obtained only by comparing the behaviour of hydrogen and deuterium (or tritium), so that this possibility belongs more properly to the subject of isotope effects.

### 3.6.3. *Tunnelling from discrete energy levels*

The treatment in the preceding sections of this chapter is based on the assumption of a continuous energy distribution (Equation 3.1). This is certainly justified for bimolecular gas reactions, where the unquantized kinetic energy of approach provides such a distribution. It also seems appropriate for bimolecular reactions in solution, since although the reacting solute species can be thought of as oscillating in a solvent cage, the energy levels of these oscillations will be close together and will be further smeared out by variations in the cage itself. However, the same picture can hardly apply strictly to intramolecular transfers of protons or hydrogen atoms, where the system is best represented by a potential energy diagram with a double minimum and well-defined energy levels, as discussed in Section 2.6. Moreover, these energy levels are quite widely spaced: for example, if the fundamental vibration frequency is 3000 cm$^{-1}$, the distance between the two lowest levels is about 36 kJ mol$^{-1}$, or 4500 $R$, where $R$ is the gas constant per mole. Since the barrier height in chemical reactions is commonly around 40 kJ mol$^{-1}$, it is clear that at ordinary experimental temperatures not more than the two lowest levels

# The application of tunnel corrections

will be involved, of which the upper will be near the top of the barrier, and often above it. Taking into account the Boltzmann distribution between the two levels the observed velocity will be

$$k = \frac{A_1 + A_2 e^{-\Delta\varepsilon/kT}}{1 + e^{-\Delta\varepsilon/kT}} \qquad (3.54)$$

where $\Delta\varepsilon$ is the energy difference between the two levels and $A_1$ and $A_2$ are their temperature-independent reaction frequencies. $A_1$ will contain a tunnelling permeability much less than unity, while the corresponding factor in $A_2$ will be much larger and will approach unity if the higher level is above the top of the barrier. For most cases of practical interest $\exp(-\Delta\varepsilon/kT) \ll 1$, when Equation 3.54 becomes

$$k = A_1 + A_2 e^{-\Delta\varepsilon/kT}. \qquad (3.55)$$

Although Equations 3.54 and 3.55 are superficially very different from those derived in Section 3.3 on the basis of a continuous energy distribution, they lead qualitatively to the same predictions, namely a rate which exceeds that predicted by classical theory and deviations from the Arrhenius equation at low temperatures. It is also found, in agreement with Section 3.5, that the observed activation energy and pre-exponential factor are both predicted to be smaller than the semi-classical values. There are of course quantitative differences between the results of the two treatments, and in particular the assumption of discrete levels leads to a much sharper change of activation energy in the neighbourhood of a temperature given by $T = \Delta\varepsilon/k \ln(A_2/A_1)$. Further, the assumption of discrete levels does not lead to an expansion in powers of $T^{-2}$ for small deviations from classical behaviour, as in Equations 3.24, 3.25, 3.38 and 3.39. It should thus be possible in principle to distinguish experimentally between the two possibilities, though this may sometimes be difficult in practice: most workers in this field have used the expressions derived for a continuous energy distribution even for where the assumption of a small number of discrete levels seems more appropriate. There is, however, one system for which Equation 3.55 is definitely to be preferred, and this is discussed in Section 6.5.

## 3.7 Summary of experimental criteria for tunnelling

It is convenient to summarize here the observable kinetic features which may serve as criteria for appreciable tunnel corrections, as shown in the preceding sections. The main consequences of tunnelling are as follows:

(a) The reaction velocity will be greater than that predicted by a semi-classical treatment. This criterion is useful only in the very few instances in which the semi-classical value can be predicted from first principles.

(b) There will be deviations from the simple Arrhenius equation in the sense that the apparent activation energy will decrease with decreasing temperature, and at sufficiently low temperatures the reaction velocity will become independent of temperature.

(c) Even when the temperature range or the experimental accuracy is insufficient to detect deviations from the Arrhenius law the observed Arrhenius parameters $(E)_A$ and $(A)_A$ will both be smaller than their semi-classical values. This criterion is useful in practice only when the tunnel correction is considerable, since only very rough values can be estimated for the semi-classical parameters $(E)_s$ and $(A)_s$.

Of these criteria the most useful results are obtained from (b), especially when experiments can be carried out over a wide range of temperatures, preferably including low ones. It will be shown in the next chapter that the study of hydrogen isotope effects provides much more useful experimental criteria of tunnelling, summarized in Section 4.3.5.

# 4 The theory of kinetic hydrogen isotope effects

## 4.1 The principles underlying kinetic isotope effects. Zero-point energies

Some of the most definite information about the part played by the tunnel effect in chemical and physical processes has come from the study of kinetic isotope effects, the theory of which is given in some detail in this chapter. The emphasis is on the comparison between the three hydrogen isotopes H, D and T, though much of the treatment has more general application. The replacement of one isotope by another can affect both equilibria and kinetics: the latter effects are by far the greater and are the only ones relevant to the tunnel effect. Kinetic isotope effects can be classified as primary, secondary and solvent effects, according to whether they involve isotopic substitution in a bond which is made or broken during reaction, in some other part of the reacting species, or in the solvent. Only primary effects are considered in this chapter. The subject is treated more extensively in a number of books and articles [112–118].

Two isotopes of the same element, or compounds containing them, were originally believed to have identical physical and chemical properties, except of course for a few properties which depend directly on molecular mass, such as density or molecular velocity. This belief appeared to be borne out by experimental evidence, since the first isotopes encountered were those of heavy elements in radioactive series, for which any isotopic differences are extremely small. The picture was drastically changed by the discovery of deuterium in 1931, since it was soon found that the replacement of hydrogen by deuterium frequently has a considerable effect on equilibrium constants and particularly on reaction rates.* A further stimulus to the study of isotope effects arose during the Second World War from the need to separate the isotopes of uranium and other heavy elements, and the fundamental theory of equilibrium isotope effects was developed during this period by Urey [120] and by Mayer and Bigeleisen [121].

---

* It is interesting to note that, at a Royal Society Discussion in 1932 [119], F. Soddy, who introduced the term 'isotope' in 1913, was reluctant to admit that deuterium could be an isotope of hydrogen, on the grounds that they differed too widely in properties.

Although our theoretical understanding of the rates of even simple reactions is still imperfect, it turns out to be a relatively simple problem to derive expressions for the effect of isotopic substitution, since many unknown factors cancel out when we are comparing two isotopic species. The principal reason for this is that *potential energy curves and surfaces are unaffected by isotopic substitution*, because interatomic and intermolecular forces depend on attractions or repulsions between the charges on electrons and nuclei, and not upon nuclear masses. This statement is true to the extent to which the motions of nuclei and electrons can be treated separately: this is the Born – Oppenheimer approximation, which may be regarded as exact in most chemical contexts.

The principles involved can be illustrated by the simple example of the homolytic or heterolytic cleavage of the two bonds X–H and X–D. The dependence of potential energy on internuclear distance will be identical for the two bonds, being represented by a curve such as that shown in Fig. 4.1. It therefore appears at first sight that the two bonds will have identical dissociation energies $E$, which for a simple dissociation process can be equated to the activation energy. The rates of dissociation of X–H and X–D should therefore differ only in virtue of their different vibration frequencies, the ratio of which for harmonic vibrations cannot exceed $2^{1/2}$. In fact the observed isotope effect $k^H/k^D$ is commonly much greater than $2^{1/2}$, so that some other factor must be involved. This is most simply expressed as the *zero-point energy*.

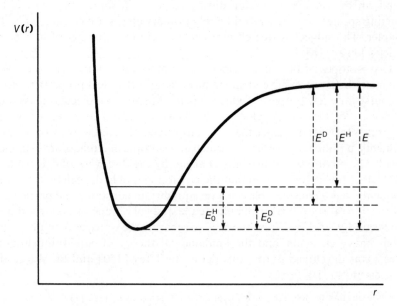

Fig. 4.1   Effect of zero-point energy in the dissociation of the bonds X – H and X – D

# The theory of kinetic hydrogen isotope effects

According to the quantum theory the lowest energy level of an oscillator is not located at the minimum of the potential energy curve, but (for a harmonic oscillator) lies above it by an amount $\frac{1}{2}hv$, where $v$ is the classical oscillator frequency. Since a single potential energy curve applies to both X–H and X–D the force constant is the same for both isotopes, but since the reduced mass is less for X–H than for X–D we shall always have $v^H > v^D$, and hence $E_0{}^H > E_0{}^D$, where $E_0$ is the zero-point energy. This is indicated in Fig. 4.1, which also shows that the two species have different dissociation energies $E^H$ and $E^D$, with $E^D > E^H$. The dissociation of the hydrogen compound thus requires a smaller activation energy than that of the deuterium compound, and should therefore be faster.

It is easily seen that zero-point energy differences can account for most of the considerable kinetic hydrogen effects observed in practice. In the expression $4\pi^2 v^2 = f/\mu$ the force constant $f$ is the same for the two isotopes, while the ratio of the reduced masses $\mu^H/\mu^D$ will be close to one half, since the hydrogen isotope is usually bound to a much heavier atom or group. In practice the observed value of $v^D/v^H$ is always somewhat greater than $2^{-1/2}$ ($=0.707$), partly because the reduced masses are somewhat greater than $m_H$ and $m_D$, and partly because of anharmonicity. An average value is $v^D/v^H = 0.741$, giving for the difference of zero-point energies*

$$E_0^H - E_0^D = \tfrac{1}{2}hv^H(1 - 0.741) = 0.130 hv^H \tag{4.1}$$

while the corresponding expression for tritium is $E_0^H - E_0^T = 0.185 hv^H$.

If the kinetic isotope effect depends solely on this difference of activation energies the value of $k^H/k^D$ is given by

$$(k^H/k^D)_s = \exp[(E_0^H - E_0^D)/kT] \tag{4.2}$$

where the subscript s denotes 'semi-classical', i.e. without allowance for tunnelling. Table 4.1 gives the calculated values for the dissociation of four common types of X–H bond. The stretching frequencies vary only slightly for a given type of bond in different species, and the values given are averages.

The values of $(k^H/k^D)_s$ in Table 4.1 indicate the rough magnitude to be expected for kinetic hydrogen isotope effects, but they relate to the unassisted cleavage of an X–H or X–D bond, a process which is not very common in

---

* In Equation 4.1 and other equations in this book the energy quantities refer to a single molecule. In tables such as Tables 3.1, 4.1, and 4.2 the same quantities are expressed *per gram-molecule*, in order to produce numbers which are familiar to the chemist, and the same procedure has sometimes been followed in the text. The same symbols, for example $E$, have therefore been used for quantities which differ by a factor $N$, the Avogadro number, and it would be more logical to use different symbols for the two sets of quantities. However, there should be little risk of confusing quantities which differ by a factor of $6 \times 10^{23}$, and I have preferred not to complicate the nomenclature further. Energies frequently appear as $E/k$ or $E/kT$, and if $E$ is given the value corresponding to one mole $k$ should of course be replaced by $Nk = R$.

Table 4.1. *Semi-classical isotope effects for the dissociation of X–H and X–D bonds*

| Bond | $v^H$ (cm$^{-1}$) | $E_0^H - E_0^D$ (kJ mol$^{-1}$) | $(k^H/k^D)_s$ at 298 K |
|---|---|---|---|
| C–H | 2900 | 4.52 | 6.2 |
| N–H | 3100 | 4.81 | 7.0 |
| O–H | 3300 | 5.10 | 7.9 |
| S–H | 2500 | 3.89 | 4.8 |

practice. A much commoner type of reaction is

$$XL + Y \rightarrow X + LY$$

where L represents H, D or T. (As it stands this scheme represents an atom transfer, but by inserting appropriate charges it can be made to represent the transfer of a proton or a hydride ion.) The appropriate energy profile for such a process is shown in Fig. 4.2. There is of course no zero-point energy associated with motion along the reaction co-ordinate, since this does not correspond to any real vibration frequency, and it appears at first sight that Equation 4.2 can still be applied. However, this reasoning ignores the fact that the transition state possesses other modes of vibration which cannot be directly shown in a diagram such as Fig. 4.2. The frequencies of these modes may be sensitive to isotopic substitution, in which case their zero-point energies will show a dependence on mass which will contribute to the kinetic isotope effect. For the three-particle transition state X——L——Y these modes are represented by (a) and (b) below,

$$\underset{\downarrow}{X}\!-\!\overset{\uparrow}{L}\!-\!\underset{\downarrow}{Y} \qquad \overset{\leftarrow}{X}\!-\!\overset{(?)}{L}\!-\!\overset{\rightarrow}{Y}$$
$$\text{(a)} \qquad\qquad\qquad \text{(b)}$$

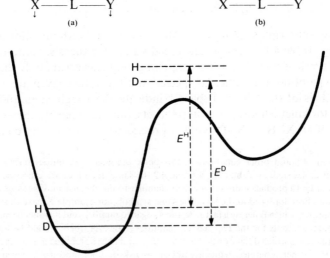

Fig. 4.2 Zero-point energies in the reaction $XL + Y \rightarrow X + LY$

# The theory of kinetic hydrogen isotope effects

The bending vibration (b) is doubly degenerate and will have a frequency strongly dependent on the mass of L, while the frequency of the 'symmetrical' vibration (b) will also be somewhat mass-dependent, since the particle L will not remain stationary unless the system is truly symmetrical.*

The zero-point energies of the transition state are indicated in Fig. 4.2 by the broken lines above the energy maximum, and it is clear that their contribution will be to reduce the isotope effect below the value which would be expected on the basis of the zero-point energies of the initial state $XL + Y$. In fact, the isotope effect will now be given by

$$(k^H/k^D)_s = \exp(\Delta E_0/kT)$$
$$\Delta E_0 = (E_0^H - E_0^D)_{XL} - (E_0^H - E_0^D)_{\ddagger} \quad (4.3)$$

where $\ddagger$ as usual refers to the transition state. Since the vibration frequencies of the transition state cannot be observed directly, Equation 4.3 does not lead to a direct prediction from spectroscopic data of kinetic isotope effects for reactions of the type

$$XL + Y \rightarrow X + YL.$$

In this it differs from Equation 4.2 for the simple dissociation of XL, and also from expressions for the corresponding isotope effect on the equilibrium constant for the reaction

$$XL + Y \rightleftharpoons X + LY$$

which involve only the zero-point energies of stable species. Comparison of Equations 4.2 and 4.3 shows that the values of $(k^H/k^D)_s$ in Table 4.1 should represent *maximum* values for reactions involving the fission of the bonds shown. A large proportion of observed isotope effects do in fact lie below these upper limits.

Apart from the effect of zero-point energy in the transition state, reactions which involve more than one elementary step may exhibit reduced isotope effects for different reasons. For example, electrophilic aromatic substitution is considered to be a two-stage process,

$$ArL + X^+ \underset{k_{-1}}{\overset{k_1}{\rightleftharpoons}} XArL^+$$

$$XArL^+ + B \xrightarrow{k_2} ArX + BL^+$$

where $X^+$ is the electrophilic reagent and B is a base, which may be the solvent. The

---

* A stable linear triatomic species has in addition an unsymmetrical stretching vibration represented by $\overset{\leftarrow}{X} \text{---} \overset{\rightarrow}{L} \text{---} \overset{\leftarrow}{Y}$, but it is of course a characteristic of the transition state that this real vibration is replaced by motion along the reaction co-ordinate, i.e. by an internal translation or vibration of imaginary frequency.

observed velocity constant for the above reaction scheme is given by

$$k = k_1 k_2 [B]/(k_{-1} + k_2[B]). \qquad (4.4)$$

Of the three velocity constants only $k_2$ will exhibit any considerable isotope effect, since the first stage of the reaction does not involve any major change in the bonding of L. It is easily seen from Equation 4.4 that this isotope effect will be observed in full only if $k_{-1} \gg k_2[B]$: when this condition is not fulfilled the isotope effect will be reduced, and will disappear if $k_{-1} \ll k_2[B]$. This behaviour has been observed in many instances [122], but will not be considered further here.

The above treatment takes no account of the zero-point energy of bending vibrations in the initial species XL. These will be present in most real systems, and arise in any model in which X is not regarded as a mass point. Bending frequencies are of course considerably lower than those of stretching vibrations and are usually more variable: for example $v_{CH}$ (bending) varies in the range 750–1450 cm$^{-1}$ in different compounds. However, since the transition state in general possesses only one fewer real vibrations than the reactants (corresponding to the appearance of internal translation along the reaction co-ordinate) it is certainly unrealistic to suppose that its formation entails the complete loss of the zero-point energy associated both with the X—L stretch and with one or more bending modes.* Moreover, it is not necessarily correct to assume that bending frequencies are lower in the transition state than in the initial state. In the system X—L—Y, although X—L and L—Y are longer than normal bonds, the particle L is now constrained on both sides and there will be a finite bending force constant even when only central forces are involved, i.e. in the absence of any preferred valency directions [123–125]. The observed bending frequency [126] for the linear symmetrical ion $HF_2^-$ is 1225 cm$^{-1}$, while calculated energy surfaces for systems such as H—H—H [127], H—H—Cl [127], H—H—CH$_3$ [128] and F$_3$C—H—CH$_3$ [129], as well as electrostatic models for proton-transfer reactions [102, 105] all predict bending frequencies in the range 500–1500 cm$^{-1}$. It thus seems likely that there is a considerable degree of cancellation between the zero-point energies of bending vibrations in the initial and transition states, in which case the values of $(k^H/k^D)_s$ in Table 4.1 can still serve as approximate estimates of the maximum isotope effects to be expected.

Interesting support for such cancellation comes from the work of Kresge and Chiang [130, 131] on the acid-catalysed hydrolysis of ethyl vinyl ether, which is known to involve rate-determining proton transfer from the acid to

---

* Strictly speaking it is not legitimate to consider the loss or gain of zero-point energy associated with individual stretching or bending vibrations, but only the overall change in zero-point energy. This is because the normal vibrational modes in general represent a combination of bending and stretching, and there is not necessarily a one-to-one correspondence between vibrations in the initial and transition states. However, the description given here is sufficiently accurate for present purposes.

carbon. Catalysis by formic acid gave $k^H/k^D = 6.8$ at 298 K, which is close to the maximum value predicted for O–H stretching alone (Table 4.1). By contrast, catalysis by hydrofluoric acid, an acid of similar strength but devoid of bending vibrations, gave $k^H/k^D = 3.3$. This is only about one quarter of the maximum value of 13.3 predicted from the known stretching frequency $v_{HF} = 4141$ cm$^{-1}$, and the difference between the behaviour of the two catalysts can be accounted for if it is assumed that the two transition states and the formic acid molecule all possess two bending frequencies of about 1100 cm$^{-1}$, which is consistent with the above estimates.

Although the contribution made by bending modes is uncertain it is of interest to give values for the maximum isotope effects to be expected when some allowance is made for these vibrations. Table 4.2 contains values calculated from Equation 4.3 on the assumption that formation of the transition state involves the complete loss of the zero-point energy of the X–L stretching vibration, and a decrease of 750 cm$^{-1}$ (when L = H) in the sum of the isotopically-dependent bending frequencies. It is assumed that $v^D/v^H = 0.741$, $v^T/v^H = 0.630$ (cf. Equation 4.1). This almost certainly exaggerates the possible effect of the bending modes, and since no allowance has been made for the stretching frequency of the transition state the values in Table 4.2 should

Table 4.2. *Maximum semi-classical isotope effects with allowance for bending vibrations*
At each temperature in (b) the first entry represents $(k^H/k^D)_s$ and the second $(k^H/k^T)_s$

(a)

| Bond | $E_0^H - E_0^D$ (kJ mol$^{-1}$) | $E_0^H - E_0^T$ (kJ mol$^{-1}$) |
|---|---|---|
| C–H | 5.66 | 8.05 |
| N–H | 5.97 | 8.50 |
| O–H | 6.29 | 8.95 |
| S–H | 5.04 | 7.17 |

(b)

| T (K) | $(k^H/k^D)_s, (k^H/k^T)_s$ | | | |
|---|---|---|---|---|
| | C–H | N–H | O–H | S–H |
| 221 | 22, 85 | 26, 109 | 31, 140 | 16, 53 |
| 243 | 17, 57 | 20, 73 | 23, 91 | 12, 37 |
| 263 | 13, 42 | 15, 52 | 18, 64 | 10, 28 |
| 298 | 10, 27 | 11, 33 | 13, 39 | 8, 19 |
| 323 | 8, 21 | 9, 25 | 10, 29 | 7, 15 |
| 363 | 6, 15 | 7, 18 | 8, 20 | 5, 11 |

represent generous estimates of the maximum isotope effects which can be accounted for by semi-classical theory. This point is important when we come to consider the interpretation of anomalously high hydrogen isotope effects in terms of the tunnel effect, and the temperatures given in Table 4.2 correspond with some of the experimental results considered in Chapter 5.

The existence and magnitude of zero-point vibrational energy can of course be deduced from any formulation of quantum-mechanical principles, such as the Schrödinger equation. It is of interest to note that it is a direct consequence of the *Heisenberg uncertainty principle*, discussed in Section 1.2 as a basis for the existence of the tunnel effect. If an oscillator could exist with no vibrational energy its state would be represented by a point at the minimum of a potential energy curve such as that shown in Fig. 4.1. This would imply that both the position and the momentum of the system could be exactly defined simultaneously, thus contravening the uncertainty principle. On the other hand, the wave function for the zero-point level yields only probability distributions for the position and the momentum, though a definite value for the energy, and the principle is satisfied. This parallelism between zero-point energy and tunnelling suggests that they will both play a part in determining kinetic isotope effects, and the role of tunnel corrections will be discussed in Section 4.3. First, however, we shall describe the more detailed semi-classical treatment based upon transition state theory.

## 4.2 The semi-classical transition state theory of kinetic isotope effects

The treatment in the last section is over-simplified in that it neglects higher vibrational states, and also the effect of mass upon the rotations and translations of the reactants. In the case of equilibria statistical mechanics enable us to formulate an essentially complete theory of isotope effects by considering the effect of mass on the translational, rotational and vibrational partition functions of the reactants and products, and this was done at an early stage in the development of the subject [120, 121]. Transition state theory enables us to apply the same methods to kinetic isotope effects, as was first shown by Bigeleisen [132]. Subsequent developments have been concerned mainly with improved methods of computation or with small corrections for anharmonicity or the failure of the Born–Oppenheimer approximation, none of which will be considered here.

### 4.2.1 *Basic expressions*
For the sake of concreteness we shall again consider the transfer reaction

$$XL + Y \rightarrow X + LY$$

though the results are easily extended to other types of reaction. The fundamental equation (Equation 3.17) then becomes

$$(k)_s = \kappa \frac{kT}{h} \frac{F_\ddagger}{F_{XL}F_Y} \exp(-\varepsilon_0/kT). \tag{4.5}$$

The energy $\varepsilon_0$ is the energy required to pass from the lowest energy state of the reactants to the lowest energy state of the transition state. When comparing two isotopic species, for example XH and XD, the expression for $(k^H/k^D)_s$ will contain an exponential factor identical with Equation 4.3, since only the zero-point energies remain when the ratio of rates is taken. This will be multiplied by the factor $F_\ddagger^H F_{XL}^D / F_\ddagger^D F_{XL}^H$, which does not occur in the simplified treatment leading to Equation 4.3.

To a good approximation each of the partition functions involved can be written as a product of translational, rotational and vibrational factors, i.e. $F = F_t F_r F_v$, and we must now investigate how each of these factors is affected by isotopic substitution. It can be safely assumed that $F_t$ and $F_r$ have their semi-classical (high-temperature) values, and the standard expressions of statistical mechanics then give, omitting constants independent of mass,

$$F_t \propto M^{3/2}, \quad F_r \propto (ABC)^{1/2}$$

$$F_v = \prod_i (1 - e^{-u_i})^{-1}, \quad u_i = h\nu_i/kT \tag{4.6}$$

in which $M$ is the total mass of the species, $A$, $B$ and $C$ are its principal moments of inertia, and the $\nu_i$ are the frequencies of its normal vibrations, assumed harmonic. Since the quantities $M$, $A$, $B$, $C$ and $u_i$ are all affected in different ways by isotopic substitution it might appear impossible to derive any simple expression for the isotope effect. However, the problem is greatly simplified by introducing the *product rules*, used for many years by spectroscopists. These relate the effects of isotopic substitution on molecular mass, moments of inertias and vibration frequencies: they are based on classical mechanics, and require only that the different degrees of freedom are independent and that the vibrations are harmonic [133]. When applied to the replacement of H by D in the stable species XL the result is

$$\left(\frac{A_{XL}^D B_{XL}^D C_{XL}^D}{A_{XL}^H B_{XL}^H C_{XL}^H}\right)^{1/2} \left(\frac{M_{XL}^D}{M_{XL}^H}\right)^{3/2} \left(\frac{m^H}{m^D}\right)^{3/2} \prod_{XL} \frac{\nu_i^H}{\nu_i^D} = 1 \tag{4.7}$$

where the product is taken over all the normal vibrations, numbering $3n - 6$ for a non-linear $n$-atomic species. Since $\nu_i^H/\nu_i^D = u_i^H/u_i^D$, combination with Equation 4.6 then gives

$$\frac{F_{XL}^H}{F_{XL}^D} = \left(\frac{m^H}{m^D}\right)^{3/2} \prod_{XL} \frac{u_i^H[1 - \exp(-u_i^D)]}{u_i^D[1 - \exp(-u_i^H)]}. \tag{4.8}$$

This procedure must be slightly modified for the transition state, which, if non-linear, has only $3n - 7$ real frequencies, plus one imaginary frequency, or internal translation, corresponding to motion along the reaction co-ordinate. Although the imaginary frequency $iv_\ddagger$ cannot be measured experimentally, the ratio $v_\ddagger^H/v_\ddagger^D$ can be replaced by $(\mu_\ddagger^D/\mu_\ddagger^H)^{1/2}$, where $\mu$ is the reduced mass for motion along the reaction co-ordinate, and the product rule becomes

$$\left(\frac{A_\ddagger^D B_\ddagger^D C_\ddagger^D}{A_\ddagger^H B_\ddagger^H C_\ddagger^H}\right)^{1/2} \left(\frac{M_\ddagger^D}{M_\ddagger^H}\right)^{3/2} \left(\frac{m_\ddagger^H}{m_\ddagger^D}\right)^{3/2} \left(\frac{\mu_\ddagger^D}{\mu_\ddagger^H}\right)^{1/2} \prod_\ddagger \frac{v_i^H}{v_i^D} = 1 \qquad (4.9)$$

where the product is taken only over the real frequencies, which are the only ones which occur in the incomplete partition function $F_\ddagger$. Combining Equations 4.6 and 4.9 we obtain

$$\frac{F_\ddagger^H}{F_\ddagger^D} = \left(\frac{\mu_\ddagger^D}{\mu_\ddagger^H}\right)^{1/2} \left(\frac{m^H}{m^D}\right)^{3/2} \prod_\ddagger \frac{u_i^H[1 - \exp(-u_i^D)]}{u_i^D[1 - \exp(-u_i^H)]}. \qquad (4.10)$$

Finally, if it can be assumed that the transmission coefficient $\kappa$ in Equation 4.5 is the same for the two isotopes, combination of Equations 4.8 and 4.10 gives for the kinetic isotope effect

$$\left(\frac{k^H}{k^D}\right)_s = \frac{F_\ddagger^H F_{XL}^D}{F_\ddagger^D F_{XL}^H} \exp(\Delta E_0/kT) = \left(\frac{\mu_\ddagger^D}{\mu_\ddagger^H}\right)^{1/2} \frac{f_\ddagger}{f_{XL}} \exp(\Delta E_0/kT) \qquad (4.11)$$

in which $\Delta E_0$ is the difference of zero-point energies, $(E_0^H - E_0^D)_{XL} - (E_0^H - E_0^D)_\ddagger$, and the function $f$ is defined by

$$f = \prod_i \frac{u_i^H[1 - \exp(-u_i^D)]}{u_i^D[1 - \exp(-u_i^H)]}. \qquad (4.12)$$

In processes where the hydrogen atom or proton is moving between much heavier atoms or groups it will be a good approximation to write $\mu_\ddagger^D/\mu_\ddagger^H = m^D/m^H$, while in other cases it can be estimated from the approximate form of the energy surface. In the semi-classical theory the transmission coefficient $\kappa$ is nothing to do with the tunnel effect, but depends on the shape of the energy surface. It has been usual to assume that $\kappa = 1$ when applying transition state theory, and the assumption that the value of $\kappa$ is independent of isotopic mass can hardly cause serious error, though it lacks strict justification.

The quantity $\Delta E_0/kT$ in Equation 4.11 can be expressed as

$$\Delta E_0/kT = \frac{1}{2}\sum_{XL}(u_i^H - u_i^D) - \frac{1}{2}\sum_\ddagger (u_i^H - u_i^D) \qquad (4.13)$$

so that the whole expression for $(k^H/k^D)_s$ can be written [apart from the factor $(\mu_\ddagger^D/\mu_\ddagger^H)^{1/2}$] in terms of the $u_i$ and hence of the vibrational frequencies only, this simplification arising from the application of the product rule. If Equation 4.13 is substituted in Equation 4.11 the latter equation can be written in the equivalent form

$$(k^H/k^D)_s = (\mu_{\ddagger}^D/\mu_{\ddagger}^H)^{1/2}(g_{\ddagger}/g_{XL}) \quad (4.14)$$

where

$$g = \prod_i \frac{u_i^H \sinh \frac{1}{2} u_i^D}{u_i^D \sinh \frac{1}{2} u_i^H} \quad (4.15)$$

and the factor involving the zero-point energies no longer appears explicitly.

Equation 4.11 or Equation 4.14 is of very general validity, subject to the applicability of transition state theory and to the following assumptions:

(a) The various degrees of freedom in the initial state and the transition state are separable, i.e., the total energy can be expressed as a sum of terms, each of which corresponds to one degree of freedom. It is this assumption which permits us to express the partition functions as products of factors corresponding to single degrees of freedom.

(b) The translational and rotational partition functions have their semi-classical (high-temperature) values. This will always be justifiable for translations, but might demand some modification at low temperatures ($T < 100$ K) for rotation of simple species containing hydrogen.

(c) All vibrations (including the imaginary vibration of the transition state) are harmonic. This condition is necessary, (i) in order to make the normal modes of vibration separable, (ii) for the validity of the expression for the vibrational partition function given in Equation 4.6, and (iii) for the validity of the product rule. Estimates have been made [134, 135] of the effects of anharmonicity on equilibrium isotope effects, which are small, but in view of our lack of detailed knowledge about transition states it is not practicable to make corresponding calculations for kinetics.

(d) The value of the transmission coefficient $\kappa$ is independent of isotopic mass. As already mentioned, this assumption cannot be strictly justified, but is intuitively reasonable.

Although the above assumptions may seem numerous they are all likely to be close to the truth in problems of chemical interest, and we may therefore regard Equation 4.11 or Equation 4.14 as substantially accurate predictions of the semi-classical theory. It should also be remembered that (as in all kinetic problems) the frequencies of the transition state cannot be determined experimentally: this gives rise to uncertainties which are greater than those associated with the assumptions which have been made.

The above treatment applies to the replacement by deuterium of a single hydrogen atom in the reactant XL. Analogous expressions are readily obtained for the isotopic replacement of several equivalent or non-equivalent atoms. No reference has so far been made to *statistical factors* or *symmetry numbers*, though these must sometimes be taken into account. For example, it is clear that even in the absence of any isotopic discrimination the reaction

$$H_2 + Y \to H + HY$$

will occur twice as fast as

$$HD + Y \to H + DY$$

and similar differences will arise whenever the isotopic substitution changes the symmetry number of the reactants or the transition state. If, as is commonly the case, a set of equivalent hydrogen atoms is completely replaced by deuterium, no symmetry numbers need be included. On the other hand, since tritium is normally used in trace amounts, it is frequently necessary to make statistical corrections when comparing tritium with the other two hydrogen isotopes. For the sake of simplicity statistical factors and symmetry numbers have been omitted from all the equations in this book.

### 4.2.2 Simplifications and model calculations

Equation 4.11 or Equation 4.14 can often be greatly simplified without much loss of accuracy. This is particularly true when we are dealing with hydrogen isotopes, and several of the expressions given in this section are not valid for heavier isotopes. We first note what happens in the classical limit, i.e. when $h \to 0$, or $T \to \infty$, or all the frequencies are very low. Under these conditions all the $u_i \to 0$, and $\sinh \frac{1}{2} u_i \to \frac{1}{2} u_i$, so that the isotope effect becomes $(k^H/k^D)_s \to (\mu_\ddagger^D/\mu_\ddagger^H)^{1/2}$. This is a purely classical effect, and since the maximum value of $\mu_\ddagger^D/\mu_\ddagger^H$ is 2, the maximum classical isotope effect is $2^{1/2}$. Much larger effects are observed in practice, showing that quantum effects must be important. It is in fact obvious that the classical limit will rarely be observed, since some of the vibrations involved are X–L stretches with $v^H = 2000 - 4000 \text{ cm}^{-1}$, for which $hv/kT = 1$ at 3000–6000 K.

Although Equations 4.11 and 4.14 contain factors for all the vibrations of the initial and transition states, in practice many of these factors will cancel out to a good approximation. This is the case for any vibration whose frequency is not changed appreciably by the isotopic substitution, and also for any vibration whose frequency remains essentially unchanged on passing from the initial to the transition state. There remain only those vibrations which are closely involved both with the isotopic substitution and with the reaction process, and this limitation greatly simplifies the problem. Thus in the reaction XL + Y which we have been considering it will be a good approximation to include only one stretching and two bending vibrations of the initial state, and two bending vibrations of the transition state.* These are all essentially stretching or bending vibrations of bonds containing hydrogen, and it is therefore usually legitimate to make the approximation $e^{-u} \ll 1$ (or

---

* The two bending vibrations can of course constitute one doubly degenerate mode. The omission of the 'symmetrical' stretching vibration of the transition state, $\overleftrightarrow{X - L_{(?)} - Y}$, can be justified because its frequency is low and will usually depend only to a minor extent upon the mass of L. This point will be considered further in Section 4.2.4.

$\sinh\frac{1}{2}u \approx \frac{1}{2}e^{u/2}$), since even for a frequency as low as 750 cm$^{-1}$ the error is only about 3% at 300 K. The expression for the isotope effect then becomes

$$\left(\frac{k^H}{k^D}\right)_s = \left(\frac{\mu_\ddagger^D}{\mu_\ddagger^H}\right)^{1/2} \left(\frac{u_1^H u_2^H}{u_1^D u_2^D}\right)_\ddagger \left(\frac{u_1^D u_2^D u_3^D}{u_1^H u_2^H u_3^H}\right)_{XL} \exp(\Delta E_0/kT). \tag{4.16}$$

Finally, if the masses of X and Y are large compared with those of the hydrogen isotopes it is a good approximation to write throughout $u^H/u^D = (\mu^D/\mu^H)^{1/2}$, giving the simple expression

$$(k^H/k^D)_s = \exp(\Delta E_0/kT) \tag{4.17}$$

identical with Equation 4.3, which was obtained on the basis of a very simple model.

If this result is accepted we can still use the values in Table 4.2 as a guide to the maximum isotope effects to be expected on the basis of the semi-classical theory. It also leads to important conclusions about the isotope effect on the Arrhenius parameters $(E)_A$ and $(A)_A$ (cf. Equations 3.30 and 3.31). In the first place, since $E_0 = (E_0^H - E_0^D)_{XL} - (E_0^H - E_0^D)_\ddagger$, the difference $(E^D)_A - (E^H)_A$ cannot exceed the difference between the zero-point energies of XH and XD, which is accessible experimentally from vibrational spectra. Secondly, the ratio $(A^H)_A/(A^D)_A$ should equal unity.

The assumption $e^{-u} \ll 1$ is certainly usually a good approximation for the initial state, and the same is probably true for the transition state, but since we have no firm knowledge of transition state frequencies it is of interest to investigate the consequences of making the opposite assumption, $u \ll 1$, for the transition state, while retaining $e^{-u} \ll 1$ for the initial state. The function $f_\ddagger$ (cf. Equation 4.12) then becomes equal to unity, and we have instead of Equation 4.16

$$\left(\frac{k^H}{k^D}\right)_s = \left(\frac{\mu_\ddagger^D}{\mu_\ddagger^H}\right)^{1/2} \left(\frac{u_1^D u_2^D u_3^D}{u_1^H u_2^H u_3^H}\right)_{XL} \exp(\Delta E_0/kT). \tag{4.18}$$

Assuming as before that $u^H/u^D = (\mu^D/\mu^H)^{1/2}$, this becomes

$$\left(\frac{k^H}{k^D}\right)_s = \frac{\mu^H}{\mu^D} \exp(\Delta E_0/kT) \approx \frac{1}{2}\exp(\Delta E_0/kT) \tag{4.19}$$

i.e., the ratio $(A_A^H/A_A^D)_s$ is close to one half. Since we have already seen that at sufficiently high temperatures $(k^H/k^D)_s$ tends to the temperature-independent value $(\mu^D/\mu^H)^{1/2} \approx 2^{1/2}$, the position can be summed up by saying that for reactions of the type

$$XL + Y \rightarrow X + LY$$

$(A_A^H/A_A^D)_s$ should always lie between $\frac{1}{2}$ and $2^{1/2}$, with a strong probability of being close to unity at ordinary temperatures.

All the conclusions arrived at so far in this section have been based on a simple model for the reaction and a number of mathematical approximations, and it is satisfactory to find that they are confirmed empirically by a large number of exact *computer experiments on model systems* of varying complexity, carried out particularly by Wolfsberg, Stern and their co-workers on the basis of the general expressions (Equation 4.11 or Equation 4.14) [136–143]. In these calculations of isotope effects assumptions are made about the geometry and force constants of the transition state. Although it is not possible to determine which detailed model is appropriate for a particular reaction, modern computing techniques make it a relatively simple matter to vary the parameters (including the temperature) over ranges which are wide enough to cover all situations which seem at all likely. Any generalizations which emerge will thus have very general validity.

The following are some of the main conclusions reached:

(a) In complicated systems no appreciable errors are caused by omitting from the calculations parts of the systems which are separated by more than two bonds from positions of isotopic substitution at which the force constants change appreciably in passing from the initial state to the transition state. This justifies the so-called *cut-off procedure*, in which a complex reacting system is replaced by a simple model, such as the X——L——Y picture used above.

(b) Certain simplifications apply only to *large primary effects*. This means in practice effects which are larger than about 2.7 at 300 K, and thus includes almost all primary hydrogen isotope effects. For such effects *zero-point energy effects are dominant*, in agreement with the conclusions reached above.

(c) The dominance of zero-point energy effects suggests that the difference in Arrhenius activation energies should not exceed the difference of zero-point energies in the initial state, and that the pre-exponential factor should be little affected by isotopic substitution. These expectations are borne out by detailed calculation. Thus Schneider and Stern [143] found, for several reaction types, a wide variation of force constants and a temperature range from 20 K to 2000 K that $(A_A^H/A_A^D)_s$ was always between 0.7 and 1.2 [0.6 and 1.4 for $(A_A^H/A_A^T)_s$], being usually much closer to unity. This is consistent with the limits of 0.5 and $2^{1/2}$ derived above on the basis of the simple model.

It is possible that in some reactions the motion of two or more protons or hydrogen atoms is concerted and synchronous, so that they move simultaneously, though direct evidence for this is hard to obtain. It is sometimes supposed that such reactions should have abnormally large isotope effects, on the grounds that the zero-point energies of more than one X–H bond are lost in the transition state [144, 145]. However, it is doubtful whether this argument can be sustained. It has been pointed out by Kreevoy [146] that the reaction co-ordinate constitutes only one normal mode, and can only replace a single vibration of the initial state: the synchronous transition state must therefore possess additional vibrational modes whose zero-point energy will tend to cancel out the zero-point energy of the extra vibration of the initial state, thus restoring a normal isotope effect. It is of course possible that the additional modes of the transition state might have low frequencies or isotopic sensitivities, and it has even been suggested [147] on the basis of a crude electrostatic model that they might have imaginary frequencies, being in fact reaction co-ordinates for the interconversion of intermediates by synchronous processes. In this event they would not cancel out any of the zero-point energy of the initial state, and abnormally large isotope effects would be anticipated. On the other hand, very general arguments [148] show that (at least to a quadratic approximation) a single point on a potential energy surface cannot represent a transition state for more than one reaction; i.e. only one mode of the transition state can have an imaginary frequency. This supports Kreevoy's argument in favour of normal isotope effects for synchronous processes, and it seems that the contrary view could only be sustained on the basis of experimental evidence which could not be explained otherwise, for example by the tunnel effect. Such evidence is lacking and may be difficult to find, since the existence of truly synchronous processes is still a matter of debate.

### 4.2.3. Comparison of all three hydrogen isotopes

The semi-classical treatment leads directly to generalizations about the relative values of $(k^H/k^D)_s$ and $(k^H/k^T)_s$. On the assumption that we need consider only three vibrations in the initial state and two in the transition state, and that zero-point energy effects are dominant, we can write

$$\frac{\ln(k^H/k^T)_s}{\ln(k^H/k^D)_s} = \frac{\sum_{XL}^{3}(v_i^H - v_i^T) - \sum_{\ddagger}^{2}(v_i^H - v_i^T)}{\sum_{XL}^{3}(v_i^H - v_i^D) - \sum_{\ddagger}^{2}(v_i^H - v_i^D)}. \tag{4.20}$$

If we further assume that $v^H : v^D : v^T = 1 : 2^{-1/2} : 3^{-1/2}$, as is reasonable for the transfer of L, L$^+$ or L$^-$ between heavy atoms, Equation 4.20 becomes

$$\frac{\ln(k^H/k^T)_s}{\ln(k^H/k^D)_s} = \frac{1 - 3^{-1/2}}{1 - 2^{-1/2}} = 1.442 \tag{4.21}$$

or alternatively

$$(k^H/k^T)_s = (k^H/k^D)_s^{1.442}. \tag{4.22}$$

Equation 4.22 is commonly known as the Swain–Schaad relation [149]. It is approximately confirmed by experiment, and is frequently used for interconverting deuterium and tritium isotope effects when only one of these has been measured.

Although the Swain–Schaad relation was originally derived from a simple model, it is again found that detailed calculations for more complex models lead to a very similar result. Thus Bigeleisen [150], using the full semiclassical equations and a more generalized model, found that the exponent in Equation 4.22 was always in the range 1.33–1.55 and usually close to 1.44. Calculations for a series of linear 4- and 5-atom transition states [151] yielded deviations of less than 2% from 1.442. Finally, very extensive computations were carried out by Stern and Vogel [142] for 180 model systems comprising five basic reaction types and a temperature range of 20–2000 K. They found that for isotope effects greater than $k^H/k^D \approx 2.7$ at 300 K the exponent was always between 1.33 and 1.58, and was close to 1.44 at ordinary temperatures for the more realistic models.

4.2.4. *The effect of transition state symmetry*
In reactions of the type XL + Y variations of either reactant will produce changes in $\Delta G$, the overall free energy change for the reaction. These will be accompanied by changes in $\Delta G^{\ddagger}$, the free energy of activation (i.e. the reaction velocity), and correlations between $\Delta G^{\ddagger}$ and $\Delta G$ represent one of the commonest types of linear free energy relationship: for proton-transfer reactions this first appeared as the Brönsted relation between catalytic power and acid–base strength [108, 152]. It is natural to enquire how kinetic isotope effects in a series of similar reactions depend on $\Delta G$, and since the transition state X——L——Y may be regarded as roughly symmetrical when $\Delta G = 0$ this problem is often described as the dependence of the isotope effect on transition state symmetry.

For proton-transfer reactions there is ample evidence for considerable variations in $k^H/k^D$ when $\Delta G$ is varied in a series of similar reactions. When $\Delta G$ is positive $k^H/k^D$ is found to increase with decrease in $\Delta G$ [153–157]. There are fewer results for negative $\Delta G$, but when such measurements have been made it is found that $k^H/k^D$ decreases as $\Delta G$ becomes more negative and passes through a maximum in the neighbourhood of $\Delta G = 0$ (symmetrical transition state) [111, 158–164]. Similar behaviour has been reported for reactions involving the transfer of hydrogen atoms [165–168], though the evidence is less abundant.

Since there are frequently large variations in $k^H/k^D$ when XL remains the same and Y is varied, their origin must be sought in variations in the properties of the transition state X——L——Y. The most commonly accepted interpretation was first put forward by Westheimer [169] and has been subsequently elaborated by a number of authors [170–172]. In its simplest form this explanation concentrates attention on the 'symmetrical' vibration $\underset{\leftarrow}{X}\underset{(?)}{\text{——L——}}\underset{\rightarrow}{Y}$ of a linear three-particle transition state, which was neglected in the simplified treatment of Section 4.2.2. If X and Y have equal masses and the force constants of the bonds X——L and Y——L are equal, so that the transition state is truly symmetrical, then the particle L will remain stationary during the symmetrical vibration, which will have a frequency independent of the mass of L. The zero-point energy of this vibration will therefore contribute nothing to the isotope effect, which (as far as stretching vibrations are concerned) will be determined by the difference between the zero-point energies of the reactants X–H and X–D. If, on the other hand, the transition state is not symmetrical, the 'symmetrical' vibration will involve some motion of L and its frequency and zero-point energy will become mass-dependent. This will reduce the isotope effect below the value found in the symmetrical case, as found experimentally (cf. Fig. 4.2).

There is no doubt that this analysis is qualitatively correct, but it seems unlikely that the contributions of this vibration to the isotope effect are large enough to explain the considerable variations in $k^H/k^D$ which are observed for moderate changes in $\Delta G$. In the neighbourhood of the transition state the potential energy of linear configurations can be written in the form

$$V = V_0 + \frac{1}{2}f_1(\Delta R_{XL})^2 + \frac{1}{2}f_2(\Delta R_{YL})^2 + f_{12}\Delta R_{XL}\Delta R_{YL} \quad (4.23)$$

where $\Delta R$ represents a departure from the value at the transition state. This expression applies both to a stable species XLY and to a transition state, the condition for the latter being $f_{12}^2 > f_1 f_2$. The standard methods of vibrational analysis then lead to the following equation for the 'symmetrical' stretching frequency $v_1$,

$$\left. \begin{array}{l} 4\pi^2 v_1^2 = \frac{1}{2}[B + (B^2 + 4C)^{1/2}] \\[6pt] B = \dfrac{f_1}{m_X} + \dfrac{f_2}{m_Y} + \dfrac{f_1 + f_2 - 2f_{12}}{m_L} \\[10pt] C = \dfrac{(f_{12}^2 - f_1 f_2)(m_X + m_L + m_Y)}{m_X m_L m_Y} \end{array} \right\} \quad (4.24)$$

analogous to Equation 3.15 for the imaginary barrier frequency. In the completely symmetrical case $f_1 = f_2 = f$, $m_X = m_Y = M$, this gives $4\pi^2 v_1^2 = (f + f_{12})/M$, i.e. $v_1$ is independent of $m_L$, as is physically obvious. By making the simplifying assumption $f_{12} = (f_1 f_2)^{1/2}$ Westheimer [169] calculated that a system in which $f_1 = 10f_2$ (or $f_2 = 10f_1$) would show an isotope effect much smaller than the maximum value given by the symmetrical system. Similar conclusions were reached in more elaborate calculations [170, 171], though the simplifying assumption $f_{12} = (f_1 f_2)^{1/2}$ was retained. Apart from the fact that a ten-fold ratio between the force constants is somewhat extreme, the assumption that $f_{12} = (f_1 f_2)^{1/2}$ is hardly realistic, since it corresponds to an energy barrier of zero curvature (cf. Equation 3.15) and hence, logically, to zero activation energy. If we make the more reasonable assumption that in the transition state the curvature of the energy surface (positive or negative) has similar magnitudes in different directions a different picture emerges [173]. It is now found that even when $f_1 = 10f_2$ the value of $v_1$ is low and only slightly affected by isotopic substitution, so that the computed isotope effect is only slightly lower than for a symmetrical transition state. The interplay of the factors $f_1$, $f_2$ and $f_{12}^2/f_1 f_2$ is well shown in an analysis by Albery [172] in which the isotope effect is calculated for independent variations of these quantities.

The above considerations apply only to the stretching frequency $v_1$ of the transition state. The bending frequency $v_3$ probably has a value between 750 and 1500 cm$^{-1}$ (cf. Section 4.1). Since it is doubly degenerate and involves major movement of L even when the transition state is symmetrical its zero-point energy will be an important factor in determining the magnitude of the isotope effect. However, it is not clear how $v_3$ will depend on the symmetry of the transition state, and in any case its effect will be largely cancelled (except when X is an atom) by the bending vibrations of the initial state XL. It is therefore usual to omit consideration of bending vibrations when interpreting the variation of $k^H/k^D$ with transition state symmetry, and in fact model calculations on the reactions

$$W-X-L+Y \rightarrow W-X+L-Y$$

and

$$W-X-L+Y-Z \rightarrow W-X+L-Y-Z$$

showed that the inclusion of bending modes had little effect on the final result [151].

Most theoretical treatments have assumed a linear configuration for the part of the transition state represented by X——L——Y. This is probably correct for the transfer of protons or hydrogen atoms, but it has been suggested [174–176] that in hydride ion transfer the transition state is triangular,

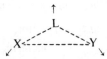

If this is so one would expect a low isotope effect even for a symmetrical transition state, since one of its real vibrations (shown by the arrows in the diagram) combines both bending and stretching and involves considerable movement of the isotope L. This conclusion is confirmed by model calculations [177], but the experimental evidence in favour of non-linear transition states is ambiguous [178]. We shall therefore omit further consideration of bending vibrations and non-linear transition states.

Returning to the linear transition state X——L——Y, the force constants $f_1$, $f_2$ and $f_{12}$ in Equation 4.23 have so far been regarded as disposable parameters. The analysis can be carried a step further [151, 177] by relating $f_1$ and $f_2$ to the *bond orders* $x$ and $1-x$ of the bonds X——L and Y——L, it being assumed that the sum of these two orders remains unity. (This is a reasonable picture when a hydrogen atom is being transferred, but less plausible for proton-transfer reactions.) There is no straightforward way of estimating the magnitude of $f_{12}$, or its variation with $x$, and arbitrary though reasonable expressions were used. As might be expected, the results show a maximum isotope effect in the neighbourhood of $x = \frac{1}{2}$, but the range of bond orders needed (roughly from $x = 0.1$ to $x = 0.9$) to account for the observed variations of $k^H/k^D$ in proton-transfer reactions seems unreasonably large. An attempt has been made [179] to estimate the bond orders for the transition states of proton-transfer reactions from the energy required to remove a proton from the reactants or products. This involves a number of questionable assumptions, but it is interesting to note that calculations for the reactions of a number of carbon acids with bases yield bond orders between 0.48 and 0.75, giving isotope effects varying only between 7.0 and 7.3, compared with the observed variation from 3.5 to 10.3.

There is no general method for correlating the structure of the transition state, or its force constants and bond orders, with measurable properties of the initial and final states, in particular the value of $\Delta G$ for the reaction. Such correlations can, however, be obtained from models, most simply from electrostatic models which are appropriate to proton-transfer reactions. The point-charge model described in Section 3.6.2 does not yield such information, since it is not consistent with any equilibrium positions for the proton in the initial and final states. This defect can be removed, somewhat artificially, by introducing a distance of closest approach of the proton to the negative charges [180], but it seems better to use a slightly more elaborate model in which the proton moves in the field of two spherical distributions of negative charge, the latter being held apart by an arbitrary repulsive potential [105]. This model yields values of $\Delta G$, which can be varied by varying the total negative charge on the spherical charge distributions, and also the vibration frequencies of the initial, final and transition states, from which the isotope effect can be calculated as function of $\Delta G$. It is found that for variations of $\Delta G$ between $-100$ and $+100$ kJ mol$^{-1}$ the magnitude of the isotope effect changes by less than 10%. This result strengthens the view that the semi-classical treatment of a three-centre model is inadequate to explain the considerable variations of $k^H/k^D$ found experimentally.

The three-centre model does not take into account a possible coupling between the extensions of the X–L or Y–L bonds and those of bonds between heavy atoms. Such coupling is frequently present, especially in proton transfers involving carbon acids, and may contribute to the variations in the isotope effect. An obvious example of coupling is in the E-2 mechanism for base-assisted $\beta$-elimination reactions, which can be formulated in the following way:

$$X-\overset{|}{C}-\overset{|}{C}-L + B \rightarrow X\text{---}\overset{|}{C}\text{===}\overset{|}{C}\text{---}L\text{---}B$$

$$\rightarrow X + \overset{|}{C}=\overset{|}{C} + LB$$

where the charges on the base B, the conjugate acid LB, and the leaving group X have been omitted. The removal of the proton is accompanied by scission of the bond C–X and conversion of the single carbon–carbon bond into a double one. It seems intuitively obvious that the degree of breaking of the C–L bond in the transition state, and hence the isotope effect, may vary considerably if the nature of X or B is changed or if different substituents are attached to the carbon atoms, even though the system as a whole must be passing through a maximum of potential energy. In terms of a more precise picture it is found [173] that the transition state shown in the above scheme now possesses a real stretching frequency (in addition to the 'symmetrical' stretch of the three-centre model) which is highly dependent on the mass of L even when L is symmetrically situated in the transition state. The same conclusions follow from detailed model calculations [181–183] in which the coupling of proton transfer with heavy-atom motion is introduced by cross terms in the potential energy function. The calculated isotope effect depends considerably on the magnitude of these cross terms, which in turn will be affected by varying the nature of either reactant. This expectation is borne out by experiment: for example, in the reaction $PhCL_2CH_2X \to PhCL = CH_2$ in presence of sodium ethoxide $k^H/k^D$ varies from 3.0 to 7.1 according to the nature of X [184], and similar variations have been reported for elimination from cis-bromocyclohexyl cyanide by a number of bases [185].

The same arguments apply to reactions in which proton transfer is accompanied not by the cleavage of another bond, but only by a change in its multiplicity. This is a common situation in slow proton transfer: for example, in reactions of the type

$$O=C-C-L+B \to {}^-O-C=C+LB$$

(where the charges on B and LB have been omitted) the coupling of C——L with C==C and C==O will produce effects entirely analogous to those envisaged for elimination reactions, and may well explain the variable isotope effects reported in [153–164] for proton abstraction from organic carbonyl and nitro compounds. Similar considerations may apply to quite simple systems: for example, the loss of a proton from $RCO_2L$ or $LCX_3$ will be accompanied by changes in the C–O distance and the XCX angle respectively, arising from changes in bond character. In terms of the semi-classical theory the coupling of proton transfer with heavy atom motion thus appears to be the most likely source of considerable variations in $k^H/k^D$ within a series, though it would not account for a maximum in the neighbourhood of $\Delta G = 0$. The tunnel effect may provide an additional cause, as discussed in Section 4.3.3.

The magnitude of the kinetic hydrogen isotope effect in a given reaction is often affected considerably by a change of solvent. Particular attention has been paid to proton abstractions by hydroxide ion in mixtures of dimethyl sulphoxide and water, for which the isotope effect often passes through a maximum at about 50% by volume of the organic component [186–191]. Since the addition of dimethyl sulphoxide to water is known to produce a large increase in the basic strength of the hydroxide ion [192] it will also affect $\Delta G$ for the reaction, and the observed variations in $k^H/k^D$ were originally attributed to changes in transition state symmetry by analogy with the interpretation of the effect of chemical changes in the reactants. However, it now seems clear that this interpretation cannot in general be applied to changes of solvent. Cox and Gibson [193] have recently investigated the free energy changes and the kinetic isotope effects for a number of simple proton-transfer reactions in mixtures of water with dimethyl sulphoxide, acetonitrile and 2, 2, 2-trifluoroethanol, but found no correlation between $k^H/k^D$ and $\Delta G$. In some instances large variations in $\Delta G$ were accompanied by almost constant values of $k^H/k^D$, while in others $\Delta G$ was almost independent of solvent composition but $k^H/k^D$ varied considerably, sometimes passing through a maximum at about 50 vol% of the organic component. There are also examples of elimination reactions in aqueous hydroxide solutions for which the isotope effect shows little variation with the content of dimethyl sulphoxide [194], while in the reaction of 4-nitrophenylnitromethane with tetramethylguanidine in a number of aprotic solvents [195, 196] $k^H/k^D$ was found to vary from 11 to 50, but bore no relation to the small variations in $\Delta G$. It must be concluded that the influence of the solvent on the magnitude of the kinetic isotope effect constitutes a complex problem, probably involving specific solvation of the reactants and the transition state. It will not be examined further here, though the role of the solvent in tunnel corrections is considered in Section 4.3.3 and in Chapters 5 and 7.

Returning to reactions in a single solvent, the expressions developed by Marcus for the dependence of reaction velocity on $\Delta G$ in a series of similar reactions can be formally used to derive the variation of the kinetic isotope effect with $\Delta G$ [197, 198]. In the simplest form of the theory the resulting expression is

$$kT\ln\frac{k^H}{k^D} = kT\ln\left(\frac{k^H}{k^D}\right)_{max} - \left(\frac{\Delta G}{4}\right)^2\left(\frac{1}{\Lambda^H} - \frac{1}{\Lambda^D}\right) \quad (4.25)$$

where $\Lambda^H$ and $\Lambda^D$ are the intrinsic barriers for the hydrogen and deuterium compounds, i.e. the values of the free energy of activation when $\Delta G = 0$. Equation 4.25 represents a parabolic dependence of the isotope effect on $\Delta G$, with a maximum effect when $\Delta G = 0$, and thus bears a formal resemblance to the results of the Westheimer treatment. However, the application of Marcus theory to isotope effects can hardly be justified. In the models on which the theory is based the only way in which $\Lambda^H$ and $\Lambda^D$ can differ is through differences in the form or dispositions of the corresponding potential energy curves. Since there is ample theoretical and experimental evidence that these curves are not affected by isotopic substitution it is not surprising that expressions such as Equation 4.25, or more elaborate versions of it, are not successful in explaining experimental results [199–201].

### 4.2.5 Conclusions of the semi-classical theory

It is convenient to summarize here the main conclusions arrived at in the preceding sections, as follows:

(a) Primary kinetic hydrogen isotope effects are determined mainly by the differences in zero-point energies in the initial and transition states. The maximum values of $(k^H/k^D)_s$ and $(k^H/k^T)_s$ expected at different temperatures for reactions involving the cleavage of various bonds are given in Table 4.2.

(b) The difference in Arrhenius activation energies $(E_A^D - E_A^H)_s$ should not exceed the difference between the zero-point energies of the initial states.

(c) The ratio of the Arrhenius pre-exponential factors $(A_A^H/A_A^D)_s$ should always lie between 0.7 and 1.2 [0.6 and 1.4 for $(A_A^H/A_A^T)$] and will usually be much closer to unity.

(d) Deuterium and tritium isotope effects for the same reaction should be related by $(k^H/k^T)_s = (k^H/k^D)^\alpha$ with $1.33 < \alpha < 1.54$, values of $\alpha$ close to 1.44 being most common.

(e) In a series of similar reactions the isotope effect is likely to have a maximum value when the reaction is thermoneutral. However, the large variations in $k^H/k^D$ observed in practice cannot be reasonably accounted for by the vibrations of a three-centre transition state, but must be attributed to coupling between proton transfer and heavy atom motion, or to some other cause.

## 4.3 Tunnel corrections to kinetic isotope effects

As shown in Chapter 1, the probability of tunnelling is closely related to the de Broglie wavelength of the particle involved. This wavelength is given by $\lambda = h/mv = h/[2m(W - V)]^{1/2}$, so that if $m$ is decreased without changing $W$ or $V$, as in isotopic comparisons, this will increase $\lambda$ and hence also the tunnelling probability. The tunnel effect thus provides an additional reason, not allowed for in semi-classical theory, for the greater reactivity normally shown by the lighter of two isotopes.

All the expressions in Chapters 2 and 3 for the permeability $G$ and the tunnel correction $Q_t$ contain the mass of the particle, and the isotope effect on these quantities is obtained by taking the ratio of the expressions with the appropriate masses inserted. These are the reduced masses for the type of motion considered, though for reactions involving hydrogen isotopes bound to heavy atoms it is frequently justifiable to use atomic masses, or some simple fraction of these masses, as discussed in Section 3.2. The observed isotope effect is then given by

$$k^H/k^D = (Q_t^H/Q_t^D)(k^H/k^D)_s \qquad (4.26)$$

where, as before, $(k^H/k^D)_s$ is the semi-classical isotope effect.

The above procedure assumes that the energy profile for tunnelling is identical for the two isotopes. This is certainly true to a very good approximation for the basic potential energy surface, but it is at least arguable [86, 87] that the zero-point energies for modes orthogonal to the reaction co-ordinate (for example the bending and real stretching vibrations of the system X——L——Y) should be added on to the basic profile along the reaction co-ordinate. Since these zero-point energies are mass-dependent and vary as the system progresses along the reaction path the resulting energy profiles will be somewhat different for the two isotopes. However, it seems unlikely that this difference has an appreciable effect on the tunnel correction. Very general arguments [202] show that in the system X——L——Y the curvature at the top of the barrier is unaffected by the inclusion of the zero-point energy of bending vibrations, suggesting that it is legitimate to use the same barrier profile for both isotopes when calculating small or moderate tunnel corrections (cf. Equation 3.23). It seems reasonable to generalize this conclusion, though strict proof is lacking.

Since the ratio $Q_t^H/Q_t^D$ (or $Q_t^H/Q_t^T$) is always greater than unity, the effect of the tunnel correction is always to increase the isotope effect above the value predicted by the semi-classical treatment. Though it is rarely possible to predict definite values of $(k^H/k^D)_s$ for a particular reaction, we have seen in the preceding sections that maximum values can be predicted with some certainty (cf. Table 4.2). If values in excess of these are observed experimentally, this constitutes good evidence for the existence of appreciable tunnel corrections, and it will be shown in Chapter 5 that such evidence of abnormally large hydrogen isotope effects is now available for a large number of reactions. However, even more valuable criteria for tunnelling can be derived from isotope effects on the activation energies and pre-exponential factors obtained from the Arrhenius equation, as described in the next section.

It has been suggested [203] that very large isotope effects may arise in the photo-dissociation of polyatomic molecules by a process analogous to tunnelling. The predicted isotope effect is of the form $\exp[(\mu_1^{1/2} - \mu_2^{1/2})K]$, where $\mu_1$ and $\mu_2$ are the reduced masses of the two isotopes and $K$ depends upon the normal modes of the initial species and of the fragments formed and upon the repulsive forces between the fragments. Calculation for the processes

$$HCN \xrightarrow{h\nu} H + CN(B^2\Sigma^+)$$
and
$$DCN \xrightarrow{h\nu} D + CN(B^2\Sigma^+)$$

predicts $k^H/k^D = 5 \times 10^{13}$! This type of effect does not appear to have been observed experimentally, though it offers obvious possibilities for separating even heavy isotopes.

### 4.3.1 *The effect of tunnelling on isotope effects for Arrhenius parameters*

The effect of tunnel corrections on Arrhenius activation energies and pre-exponential factors, as defined operationally by Equations 3.31 and 3.32, has been treated in Section 3.5, which contains explicit expressions for parabolic barriers. The isotope effects on these quantities are again obtained by evaluating the corrections for different values of the particle mass. For the small or moderate tunnel corrections which are usually appropriate to chemical reactions, the difference between the actual and semi-classical Arrhenius parameters will always be greater for the lighter isotope, leading at once to the following predictions,

$$E_A^D - E_A^H > (E_A^D - E_A^H)_s, \quad A_A^D/A_A^H > (A_A^D/A_A^H)_s. \tag{4.27}$$

These predictions are of great value in the experimental detection of tunnelling. Because of our ignorance of the vibration frequencies (and hence zero-point energies) of the transition state it is rarely possible to give a firm value for $(E_A^D - E_A^H)_s$, but it cannot exceed $E_0^H - E_0^D$, the isotopic difference in the zero-point energies of the initial state (cf. Equation 4.3), which is directly accessible from vibrational spectra. The observation of values of $E_A^D - E_A^H$ which appreciably exceed $E_0^H - E_0^D$, or the corresponding observation for tritium isotope effects, thus constitutes good experimental evidence for a sizeable tunnel correction, and many such cases are quoted in Chapter 5. The position is even more favourable for the Arrhenius pre-exponential factors, since we have seen in Section 4.2.2 that the application of semi-classical theory to a wide range of models leads to the conclusion that $(A_A^D/A_A^H)_s$ will always lie between 0.8 and 1.4, being usually closer to unity: thus the observation of values of $A_A^D/A_A^H$ greater than 1.4 (or 1.8 for $A_A^T/A_A^H$) provides an excellent test for tunnelling. This criterion is particularly valuable since it does not involve any knowledge of the reaction concerned, and may be applicable when (because of the zero-point energy of the transition state) $E_A^D - E_A^H$ does not exceed $E_0^H - E_0^D$ and $k^H/k^D$ is less than the maximum value expected for $(k^H/k^D)_s$.

The position is different if the extent of tunnelling is large, as will be the case for any hydrogen-transfer process at sufficiently low temperatures. This is readily seen by considering what happens in the limit of very low temperatures, when reaction takes place solely by tunnelling from the state of lowest energy. Both $k^H$ and $k^D$ then become independent of temperature, so that the observed activation energies, and their difference, tend to zero. The ratio of the pre-exponential factors becomes equal to the ratio of permeabilities, $G^D/G^H$, which is of course much less than unity.* It is therefore clear that over some range of low temperatures we shall have

$$E_A^D - E_A^H < (E_A^D - E_A^H)_s, \quad A_A^D/A_A^H < (A_A^D/A_A^H)_s \tag{4.28}$$

which is just the reverse of the predictions in Equation 4.27 for moderate tunnel corrections. The same conclusions can be expressed more quantitatively in terms of Equations 3.40 and 3.41, derived for a large degree of tunnelling through a parabolic barrier provided that $\alpha$ and $\alpha - \beta$ are sufficiently large.

If a given reaction is studied over a sufficiently wide range of temperature, both the conditions in Equation 4.27 and in Equation 4.28 should be realisable at high and low temperatures respectively. Since the calculated values of $(E_A^D - E_A^H)_s$ and $(A_A^D/A_A^H)_s$ vary only slightly with temperature [140], plots of $E_A^D - E_A^H$ or $A_A^D/A_A^H$ against temperature will pass through a maximum, and there will be an intermediate temperature range in which $E_A^D - E_A^H$ and

---

* There are actually two reasons why $A^D/A^H \ll 1$ under these extreme conditions, firstly because for a given energy $G^D/G^H \ll 1$ on account of the occurrence of the isotopic mass in all the expressions for $G$, and secondly because the higher zero-point energy of $X-H$ gives it an advantage over $X-D$ in tunnelling through a given barrier.

$A_A^D/A_A^H$ are close to their semi-classical values, although the degree of tunnelling is large and $k^H/k^D$ is abnormally high. Thus although markedly non-classical values of the isotope effect on the Arrhenius parameters indicate the presence of tunnel corrections of considerable magnitude, the observations of values close to the semi-classical ones does not exclude a large degree of tunnelling provided that the observed isotope effect is large.

This behaviour is shown clearly by the model calculations of Stern and Weston [204] for several different reaction types and a temperature range 20–2000 K. The semi-classical quantities were calculated by using the full expressions such as Equation 4.11 or Equation 4.14, and the tunnel corrections computed numerically for an Eckart barrier (Equation 2.46) having a curvature at the top equal to that calculated from the vibrational analysis of the model. The plots of $\ln(Q_t^H/Q_t^D)$ against $1/T$ are in all cases sigmoid curves, corresponding to the behaviour of $E_A^D - E_A^H$ and $A_A^D/A_A^H$ described above. The models suggest that the changeover from the condition in Equation 4.27 to that in Equation 4.28 may well occur at readily accessible temperatures. In practice it may be difficult to cover a sufficient temperature range for both isotopes, since the isotope effect will be very large in the region of interest, but some indications of such behaviour are mentioned in the review of experimental evidence in Chapter 5.

### 4.3.2 Tunnel corrections for all three hydrogen isotopes

As shown in Section 4.2.3, semi-classical theory predicts a relation between deuterium and tritium isotope effects of the form $\ln(k^H/k^T)_s = y\ln(k^H/k^D)_s$, where $y = 1.442$ in the simplest form of the theory, but varies between 1.33 and 1.55 in a more elaborate treatment. It might be expected that the introduction of large tunnel corrections would lead to values of $y$ outside this range, but closer consideration [205] shows that this is unlikely to be the case. If we define an exponent $x$ ($> 1$) by the relation $Q_t^H/Q_t^T = (Q_t^H/Q_t^D)^x$, then since $k = Q_t k_s$ for each isotope we find

$$y' - y = \frac{x - y}{1 + [\ln(k_s^H/k_s^D)/\ln(Q_t^H/Q_t^H)]} \quad (4.29)$$

where $y'$ is the observed value of $y$. This equation shows that $y'$ will differ appreciably from $y$ only when $x$ and $y$ are significantly different and when $k_s^H/k_s^D$ is as close as possible to its minimum value of unity, i.e. when the isotope effect is mainly due to the tunnel correction. It seems unlikely that these conditions will be fulfilled, and this conclusion is borne out by detailed model calculations for a variety of reaction types [206]. These calculations show that the tunnel corrections have little effect on the value of the exponent $y$, and will not lead to values outside the semi-classical range 1.33–1.55 for any reasonable models. A comparison of tritium and deuterium isotope effects is therefore unlikely to be directly diagnostic of tunnel effects, and in fact normal values of $y$ have been observed in reactions which by other criteria involve large tunnel corrections.* However, such a comparison is of course valuable as a quantitative test of any theory of isotope effects, irrespective of whether tunnel corrections are involved.

### 4.3.3 Tunnelling and the dependence of isotope effects on transition state symmetry

Section 4.2.4 discusses the semi-classical interpretation of the observed dependence of kinetic hydrogen isotope effects on transition state symmetry, the latter being usually measured by the overall free energy change of the process. It was concluded that the usual interpretation, originally due to Westheimer, in terms of the 'symmetrical' vibrations of a three-centre transition state, is probably inadequate to explain the observed variations, but that this might be accounted for by the coupling of proton or hydrogen-atom transfer with other types of motion. The tunnel correction may also contribute to the variations of isotope effect within a series, and this possibility is considered briefly here.

It was shown in Section 3.6.2 that for a fixed particle mass the tunnel correction may have a maximum value when the process is thermoneutral, i.e. when the transition state is symmetrical. The same will be true for the ratio $Q_t^H/Q_t^D$, so that the tunnel effect could in principle account for the fact that $k^H/k^D$ sometimes passes through a maximum when $\Delta G$ is close to zero. This possibility has been investigated for a model system in which the proton moves in the field of two spherical distributions of negative charge [105]. Fig. 4.3 illustrates typical plots of $Q_t^H/Q_t^D$ against $\Delta G$, which show that the tunnel effect could well be the cause of the observed maxima. In terms of the same model the variations in $k^H/k^D$ due to the real vibrations of the transition state amount to less than 10%, so that the plots of $k^H/k^D$ against $\Delta G$ have essentially the same form as those of $Q_t^H/Q_t^D$. As mentioned in Section 4.2.4, the coupling of proton transfer with the motion of heavy atoms in the system can produce considerable variations in $(k^H/k^D)_s$, but there is no reason why this should lead to a maximum in the plot of isotope effect against $\Delta G$.† These considerations

---

* One highly abnormal value ($y = 1.12$) has been reported for the abstraction of protons from acetone by hydroxide ions [207]. The deuterium effect was determined by using bromine as a scavenger for the anions of acetone and acetone-$d_6$, and the tritium effect from the detritiation rate of acetone containing trace quantities of tritium. However, mass-spectrometric measurements of the rate of dedeuteration of acetone [208] gave results differing considerably from those derived from the bromination experiments, so that the abnormal exponent originally reported is probably spurious.

† Intuitively one would expect the tunnel correction to be reduced by such coupling because of the resulting increase in the reduced mass for motion along the reaction co-ordinate. However, detailed calculations [209] show that the reverse is the case: the reason is that the increased reduced mass is more than compensated by an increase in the curvature of the barrier.

# The theory of kinetic hydrogen isotope effects

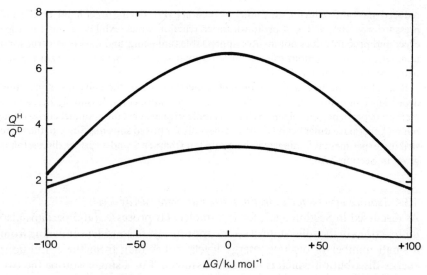

Fig. 4.3 Dependence of tunnel correction to isotope effect on free energy change calculated from an electrostatic model (from [105])

suggest that tunnel corrections might be the main factor responsible for the observed maxima, and the same conclusion has been reached on the basis of a different picture [210], in which the process is assumed to take place by tunnelling from the lowest vibrational level, the activation energy being mainly associated with solvent reorganization. However, opinions differ as to the importance of tunnelling for the variation of $k^H/k^D$ in a series, and such observations cannot at present be used as a reliable guide to the magnitude of tunnel corrections.

Tunnel corrections are probably also implicated in the *effect of the solvent* on the magnitude of isotope effects, especially for proton-transfer processes. These will always involve a redistribution of charge, which in a polar solvent will be accompanied by a re-orientation of solvent molecules. The result will be to increase the reduced mass for motion along the reaction co-ordinate and hence to reduce the tunnel correction and the isotope effect. This explanation has been invoked to explain the fact that the very large isotope effects observed for some proton transfers in solvents of low polarity are greatly reduced in polar solvents [195, 196], and the effect of a polar solvent on the tunnel correction also emerges from model calculations [209]. A similar interpretation has been given for the *effect of pressure* on some hydrogen isotope effects. The reaction between leuco crystal violet and chloranil is believed to involve hydride ion transfer, and there is independent evidence [211] that a considerable tunnel correction is involved. An increases of pressure from 1 bar to 2000 bars decreases $k^H/k^D$ in acetonitrile from 11.5 to 7.5, and it is suggested that at high pressures the solvent molecules are forced into closer coupling with the hydride ion, thus increasing the reduced mass and reducing the tunnel correction [212]. On the other hand, no effect of pressure was observed in a proton-transfer reaction which exhibits a small isotope effect and probably does not involve appreciable tunnelling, and the same is true for a number of hydrogen-atom transfers in non-polar solvents, for which solvent participation should be minimal [213].

It must be admitted that our present understanding of the solvent (or pressure) dependence of hydrogen isotope effects is far from satisfactory. It is likely that a very marked dependence on solvent polarity is indicative of substantial tunnel corrections, but no general treatment can be given, especially for mixed solvents. Some of the more striking experimental findings will be quoted in Chapter 5, and a further discussion is given in Section 7.2.4.

### 4.3.4 *Isotope effects for tunnelling from discrete energy levels*

As discussed in Section 3.6.3, for intramolecular processes (and perhaps also for reactions in the solid state) it seems appropriate to consider tunnelling from a small number of discrete energy levels rather than from the continuous energy distribution which is normally assumed. For a single isotope the two types of treatment lead to observable results which are qualitatively similar, and the same will often be true for the effect of tunnelling on isotope effects. This can be seen by reference to Equation 3.55. At fairly high temperatures the second term will be dominant, and if $A_2^H$ and $A_2^D$ have similar values the isotope effect will be determined by the difference between the energy intervals $\Delta\varepsilon^H$ and $\Delta\varepsilon^D$, and will obey the Arrhenius equation. At low temperatures the first term becomes dominant, leading to a large temperature-independent isotope effect given by $A_1^H/A_1^D$, the ratio of the barrier permeabilities. At intermediate temperatures there will be deviations from the Arrhenius equation and the ratio of the observed pre-exponential factors $A_A^H/A_A^D$ may be either greater or less than unity.

On the other hand, the effect of isotopic mass on the spacing of the energy levels may sometimes lead to behaviour which is peculiar to the particular system being considered. Thus it may well happen that the first excited vibrational levels for hydrogen and deuterium are respectively above and below the maximum of a double-minimum potential. In this event we should have $A_2^H \gg A_2^D$, and it might be necessary to include the second excited level for deuterium, but not for hydrogen. This would lead to a complicated situation, especially if all three isotopes of hydrogen are being considered. These possibilities do not appear to have been realised experimentally for chemical reactions, but tunnelling between discrete levels has been used to account for a number of apparent anomalies in the diffusion of the hydrogen isotopes in metals, notably the fact that in palladium the order of diffusion coefficients is D > H > T [214, 215], and that in tantalum the activation energy for diffusion changes fairly sharply at temperatures which are different for hydrogen and deuterium [216].

## The theory of kinetic hydrogen isotope effects

### 4.3.5 Experimental criteria for tunnelling in kinetic hydrogen isotope effects

It is convenient to collect here the observable features of kinetic hydrogen isotope effects which may be used as criteria for appreciable tunnel corrections, as shown in the preceding sections. The main consequences of tunnelling are as follows:

(a) The isotope effect will exceed the value calculated by semi-classical theory. Maximum values of the latter can be estimated from spectroscopic data (cf. Table 4.2).

(b) For moderate tunnel corrections the Arrhenius parameters $E_A^D - E_A^H$ and $A_A^D/A_A^H$ will exceed their semi-classical values. The semi-classical value $(E_A^D)_s - (E_A^H)_s$ cannot exceed the isotopic difference in zero-point energies in the initial state, which can be estimated from spectroscopic frequencies, while by model calculations $(A_A^D)_s/(A_A^H)_s$ can be fixed within narrow limits which are independent of the details of the model.

(c) When the tunnel correction is very large (and hence always at a sufficiently low temperature) the above relations between observed and semi-classical Arrhenius parameters are reversed. For any reaction there will therefore be a temperature range in which the dependence of the Arrhenius parameters on isotopic mass is close to that predicted by semi-classical theory, although the tunnel correction and the isotope effect are both large.

(d) Tunnel corrections are not likely to cause significant deviations from the semi-classical relation $k^H/k^T = (k^H/k^D)^y$, with $y$ between 1.33 and 1.55.

(e) Tunnel corrections are probably often implicated in the variation of $k^H/k^D$ within a series of similar reactions, and also in the influence of solvent or pressure upon the isotope effect, though it is not usually possible to separate the effect of tunnelling from those of other factors.

Comparison of points (a), (b) and (c) above with the corresponding criteria for measurements involving a single isotope (Section 3.7) shows the advantage gained by studying isotope effects, since in place of semi-classical quantities such as $k_s$, $(E_A)_s$ and $(A_A)_s$, which are rarely available, we are now able to make comparisons with $(k^H/k^D)_s$, $(E_A^D - E_A^H)_s$ or $(A_A^D/A_A^H)_s$, which can be assigned values or limits with some confidence.

Another aspect of the experimental consequences of tunnelling has been treated by Simonyi and Mayer [217] who consider the effect on the tunnel correction of varying the width of a barrier of given type and height. They find that the values of $k^H/k^D$, $E_A^D - E_A^H$ and $A_A^D/A_A^H$ all pass through a maximum at a particular barrier width in the range 35–70 pm. These maximum values at 300 K have been tabulated for parabolic and Eckart barriers, together with the corresponding values for tritium, and are based on the assumption of a continuous Boltzmann energy distribution. None of the experimental values reported in Chapter 5 exceed the predicted maxima, which may prove useful in detecting spurious experimental results. It is interesting to note that calculations on the basis of an electrostatic model of proton transfer [105] also predict a maximum tunnel correction for a particular distance between reactants.

# 5 Experimental evidence for tunnelling in chemical reactions

## 5.1 Introduction

The experimental criteria which may be used to detect appreciable tunnel corrections in chemical reactions have been summarized in Sections 3.7 and 4.3.5.* The first clear experimental evidence for such corrections was published in 1956 in a study of the Arrhenius parameters in the abstraction of protons or deuterons from 2-ethoxycarbonylcyclopentanone [218], and since then an increasing volume of evidence has been obtained for a wide variety of reactions. A summary of such evidence was published in 1974 [117], and several striking examples have come to light subsequently.

The experimental material comprises deviations from the Arrhenius equation, abnormally large kinetic isotope effects, and abnormal isotope effects upon the Arrhenius parameters. Since more than one of these features has often been studied for a given process, it is convenient to consider separately four broad classes of reaction. Section 5.2 deals with hydrogen atom transfers which have been studied down to low temperatures, often in solids or glasses, thus giving rise to striking effects. Section 5.3 describes some of the evidence relating to hydrogen atom transfers in very simple gas reactions, where there is some prospect of a more detailed theoretical treatment. Section 5.4 attempts a general survey of relevant results for kinetic hydrogen isotope effects in solution: these relate largely to proton-transfer (acid–base) reactions, but also include reactions involving the transfer of hydrogen atoms or hydride ions. In this category the results for a single reaction are frequently not very extensive or accurate, but the cumulative picture is quite convincing and is discussed in Section 5.5. Finally, Section 5.6 describes the few instances in which observations at extremely low temperatures have been interpreted in terms of tunnelling by entities much heavier than hydrogen, and also mentions the interesting possibilities offered by the chemistry of muonium, which may be regarded as an isotope of hydrogen of mass 0.113.

* It should perhaps be emphasized that there is no question of testing experimentally the fundamental concepts of the tunnel effect, which are firmly based in the quantum theory, but only of establishing how large an effect is produced in practice by tunnel corrections.

## 5.2 Hydrogen atom transfers at low temperatures

The experiments described in this section have all been carried out during the last decade, and since they provide some of the most striking demonstrations of the reality of the tunnel effect they will be described in some detail. A common feature of this work is that atoms or free radicals are produced in liquids, solids or glasses by radiation. Electron spin resonance (e.s.r.) spectroscopy is then used to observe the disappearance of the atoms or radicals, or the appearance of new radical species, and in some instances both processes could be followed. Since the details of an e.s.r. spectrum are sensitive to small changes in structure or environment it is also possible to follow the isomerization of organic radicals, the interconversion of radical pairs and isolated radicals, and even the transformation of one diradical pair into another which is chemically identical but is situated differently in a crystal.*

An extensive series of investigations of this kind has been carried out by Ffrancon Williams and his co-workers [219–224] on reactions of the type

$$CH_3{}^{\cdot} + CH_3X \rightarrow CH_4 + {}^{\cdot}CH_2X$$

where X is CN, NC, or OH. The first investigations dealt with acetonitrile. The action of $\gamma$-rays on a crystal of this substance produces colour centres, which can be identified from their e.s.r. spectrum [225, 226] as solvated electrons $e^{-}(CH_3CN)_2$. If the crystal is then irradiated with visible light the process

$$e^{-}(CH_3CN)_2 \rightarrow CH_3{}^{\cdot} + CN^{-} + CH_3CN$$

known as photo-bleaching, takes place. The methyl radicals thus produced disappear at a measurable rate at temperatures below about 120 K, to a small extent by the reversal of the photo-bleaching process, which can be monitored spectrophotometrically, but mainly by hydrogen abstraction from a neighbouring acetonitrile molecule,

$$CH_3{}^{\cdot} + CH_3CN \rightarrow CH_4 + {}^{\cdot}CH_2CN.$$

The rate of the last process can be measured by observing either the disappearance of the e.s.r. spectrum of ${}^{\cdot}CH_3$ or the appearance of that of ${}^{\cdot}CH_2CN$. These two observations give concordant values for the velocity constant, and there is also agreement between values obtained after shutting off the visible light and under conditions of constant illumination. There is thus little doubt about the nature of the reaction being observed or the magnitude of its velocity constant.

---

* Valuable information about tunnelling in hydrogen-transfer processes has also been obtained by nuclear magnetic resonance (n.m.r.) spectroscopy. However, the principle involved is not the same as in the e.s.r. studies described here. The n.m.r. spectra are not used as an analytical tool for monitoring changes in the concentrations of different species, but instead the form of the spectrum is used to deduce the life-time of the species concerned and hence the rate at which they are interconverted. This type of evidence will therefore be deferred until Chapter 6, where it will be discussed together with other spectroscopic phenomena involving tunnelling.

General experience of hydrogen abstraction reactions in the gas phase would suggest that in the absence of tunnelling no measurable reaction should occur at 77 K and 87 K, the temperatures first studied [219]. Later measurements [220] extended the range to 69–112 K and produced a strongly curved Arrhenius plot, in which the apparent activation energy varies from 3 to 10 kJ mol$^{-1}$. Since the observed activation energy for the gas reaction $CD_3$ + $CH_3CN$ at 373–573 K is $42 \pm 2$ kJ mol$^{-1}$ [227] this is strongly indicative of tunnelling. A quantitative treatment becomes possible with the aid of one further assumption, namely that the classical pre-exponential factor ($J_0$ in Equation 3.3) is equal to the frequency of C–H stretching vibrations in solid acetonitrile, which experiment shows to be $9 \times 10^{13}$ s$^{-1}$, differing little from the frequency in the gas phase [228, 229]. The observed velocity constant is then given by

$$k = 9 \times 10^{13} Q_t \exp(-V_0/kT) \tag{5.1}$$

where $Q_t$ is the tunnel correction defined by Equation 3.4 and $V_0$ is the barrier height. Any type of barrier can be characterized by the value of $V_0$ and a width parameter $b$, and it is then possible to test whether values of $V_0$ and $b$ can be found which will reproduce the observed velocity constant over the whole temperature range. Such computations were carried out by Le Roy, Sprague and Williams [220] for symmetrical truncated parabolic barriers, Eckart barriers (Equation 2.46) and Gaussian barriers (Equation 2.68). They found that an exact quantitative fit could be obtained with the Gaussian barrier, while the Eckart barrier was slightly less satisfactory and the parabolic barrier failed to give quantitative agreement. The best values of the parameters of the Gaussian barrier were $V_0 = 44$ kJ mol$^{-1}$, $b = 64$ pm. The former is substantially identical with the observed activation energy for the high-temperature gas reaction, and the latter is physically reasonable for the half-width of the barrier. Most of the reaction must involve systems with energies near the base of the barrier, and it is understandable that good agreement could not be obtained with a truncated parabolic barrier, which is obviously unrealistic in this region.

The values of the tunnel correction $Q_t$ in the experimental temperature range are very large, being between $10^5$ and $10^{15}$, so that it is entirely reasonable to describe the reaction as taking place by a tunnelling mechanism under these conditions. The theoretical treatment predicts very large hydrogen isotope effects, the calculated value of $k^H/k^D$ ranging from about $10^3$ at 120 K to $10^6$ at 77 K. In practice it is only possible to establish lower limits for $k^H/k^D$, since in $CD_3CN$ the abstraction reaction cannot be detected experimentally, the $CD_3$ radicals disappearing preferentially by other routes, such as the reversal of the photo-bleaching process. In the original measurements [219] it was only possible to establish a limit of $k^H/k^D > 140$ at 87 K, which is less than the semi-classical maximum value of 500 calculated from the zero-

## Experimental evidence for tunnelling

point energies of the C–H and C–D vibrations. However, it was subsequently reported [222] that $k^H/k^D > 2000$ at 87 K, while later work by Sprague [224], using refined techniques, showed that $k^H/k^D > 28\,000$ at 77 K, which is well above the semi-classical maximum of 1500 at this temperature.

Very similar results were obtained in a study of crystals of methyl isocyanide [221], in which $\gamma$-irradiation gives rise directly to methyl radicals. The disappearance of $CH_3\cdot$ and the simultaneous appearance of $\cdot CH_2NC$ were studied over the range 77–125 K, concordant velocity constants being obtained. The Arrhenius plot is strongly curved, yielding apparent activation energies which decrease from 20 to 6 kJ mol$^{-1}$ with decreasing temperature. The reaction $CD_3\cdot + CD_3NC \rightarrow CD_4 + \cdot CD_2NC$ could not be detected, and it was concluded that $k^H/k^D > 1100$ at 110 K, compared with a semi-classical maximum of 150. All these results are consistent with an interpretation in terms of tunnelling, though no quantitative treatment was attempted. At 120 K $CD_3\cdot$ radicals disappear slowly without the formation of $\cdot CD_2NC$, probably by recombination to give $C_2D_6$, while under the same conditions $CH_3\cdot$ radicals react with $CH_3NC$ to give exclusively $CH_4 + \cdot CH_2NC$. This kind of behaviour has been termed an 'all-or-nothing' isotope effect, since the hydrogen and deuterium compounds give different products. Such situations will arise whenever there are two competitive pathways, only one of which has a very large isotope effect: they may conceivably be of preparative value.

Tunnelling is also prominent in the reaction

$$\cdot CH_3 + CH_3OD \rightarrow CH_4 + \cdot CH_2OD$$

in methanol-$d$ glasses [222, 223].* In this system the methyl radicals were produced either by photolysis of methyl iodide, or by photo-ionization of $p$-$C_6H_4(NMe_2)_2$ followed by dissociative electron transfer to methyl chloride. The abstraction reaction was again followed by e.s.r. monitoring of both $\cdot CH_3$ and $\cdot CH_2OD$. The original measurements [222] at 67 K and 77 K gave an activation energy of 4 kJ mol$^{-1}$, much less than the value of $34 \pm 4$ kJ mol$^{-1}$ reported for the same reaction in the gas phase at 376–492 K [230], and an isotope effect $k^H/k^D > 1000$ at 77 K. Later work [223] covered the very wide temperature range 10–90 K. The Arrhenius plot is markedly curved, and below about 40 K there is no detectable variation of rate with temperature. It is stated that the results above 40 K are consistent with a Gaussian barrier of height 35 kJ mol$^{-1}$ and half-width $b = 69$ pm, but that this predicts too large a temperature dependence below 40 K. The last discrepancy may arise because the computations assume a continuous Boltzmann energy distribution, while at the low temperatures involved it is likely that tunnelling takes place from a small number of discrete levels, as discussed in Section 3.6.

---

* $CH_3OD$ was used instead of $CH_3OH$ in order to avoid the broadening of the e.s.r. spectrum of $\cdot CH_2OH$ by interaction with the hydroxyl proton.

Observations of e.s.r. spectra have also been used by K.U. Ingold and his collaborators [231–233] as an analytical tool for studying the isomerization of sterically-hindered aryl radicals. Although simple aryl radicals have very short lifetimes, usually decaying by abstraction from the solvent, they can be greatly stabilized by introducing bulky substituents. In suitable systems it is then possible to use e.s.r. spectroscopy to follow the isomerization of the initial radical into a more stable one. The best studied example is the isomerization of 2,4,6-tri-*t*-butylphenyl into 3,5-di-*t*-butylneophyl, i.e.

I–H

[structure: Me$_3$C and CMe$_3$ at 2,6-positions of phenyl radical with CMe$_3$ at 4-position → Me$_3$C and CMe$_2\dot{C}H_2$ at 2,6-positions with CMe$_3$ at 4-position]

together with the corresponding reaction for the completely deuterated species (I–D): the process thus involves the intramolecular transfer of a hydrogen or deuterium atom. The initial radicals can be generated in solution by the photolysis of *t*-butyl 2,4,6-tri-*t*-butylperbenzoate, or, less directly, by the reaction of 2,4,6-tri-*t*-butylbromobenzene with the radical $R_3Sn^\cdot$ generated photochemically *in situ* from $Sn_2R_6$. The rate of isomerization was measured by monitoring the disappearance of the initial radical either after cutting off the illumination or under conditions of steady illumination. The rates were found to be the same in several solvents, and at low temperatures the same results were obtained in liquid solvents and in solid matrices, thus providing good evidence that the observed process is intramolecular.

The temperature range studied was 28–247 K for the radical I–H and 120–300 K for I–D. The Arrhenius plots for both species are shown in Fig. 5.1, which include values obtained in several solvents and in a solid neopentane matrix. Both plots are non-linear, and in the case of I–H the reaction velocity becomes virtually independent of temperature below about 40 K. In the temperature range common to both species the isotope effect $k^H/k^D$ varies from 80 at 243 K to 13 000 at 123 K: the corresponding semi-classical maximum values are 17 and 260, respectively.

These findings are clearly indicative of tunnelling, and the authors have shown [232, 233] that their results are in fair quantitative agreement with tunnelling corrections calculated on the basis of various types of barrier. In addition to parabolic, Eckart and Gaussian barriers they have included a Lorentzian barrier defined by the equation

$$V(x) = V_0\left(1 + \frac{x^2}{b^2}\right)^{-1}. \tag{5.2}$$

*Experimental evidence for tunnelling* 111

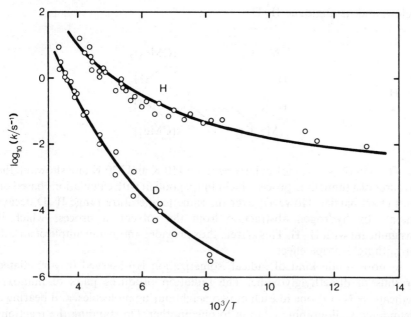

Fig. 5.1 Arrhenius plot for the isomerization of 2,4,6,-tri-*t*-butylphenyl and its completely deuterated analogue (data from [231–233])

Since the experimental results cover a very wide range of temperatures it is not necessary to make any assumption about the semi-classical pre-exponential factor $(A)_s$, which, together with $V_0$ and $b$, can be derived from the results. Of the barriers investigated the Gaussian and Eckart types gave the best fit, and the curves drawn in Fig. 5.1 have been calculated from the latter. It is satisfactory to find that the resulting values of $(A)_s$, $V_0$ and $b$ are all physically reasonable, and that $V_0^D - V_0^H$ is always less than the difference between the initial zero-point energies. However, as emphasized by the authors, in view of the simplified nature of the model it is unwise to attach too much importance to the differences between barriers of different types or to the values of the parameters derived from them. The rate at 28 K predicted by the Eckart or Gaussian barriers is considerably less than that actually observed, and this may be due to the neglect of discrete energy levels, as already suggested in connection with low-temperature abstraction from methanol by methyl radicals [223].

Ingold and his collaborators also investigated the isomerization of the 2,4,6-tri (1'-adamantyl)phenyl radical between 106 K and 204 K, obtaining a non-linear Arrhenius plot which is well reproduced by calculations with a Gaussian or Eckart barrier. The deuterated analogue of this radical was not investigated, but interesting results were obtained with the radical II-H, shown below, and its analogue II–D.

II–H

(structure: bicyclic aromatic with (Me₃C)₂ substituents on left ring positions and (CMe₃)₂ on right ring positions, H₂ groups, and a central radical •O)

II–H isomerizes intramolecularly between 110 K and 320 K and shows all the features of a tunnelling process, including a good fit with calculations based on an Eckart barrier. However, over the same temperature range II–D decays mainly by hydrogen abstraction from the solvent, a process which is insignificant with II–H. This system thus provides another example of an 'all-or-nothing' isotope effect.

A more subtle kind of radical isomerization is observed in $\gamma$-irradiated crystals of dimethylglyoxime. The radiation produces pairs of iminoxy-radicals, $=N-O^\bullet$; one in each of two neighbouring molecules, and bearing a definite crystallographic relation to one another. On standing the reaction

$$=N-O^\bullet + HO-N= \rightarrow =N-OH + {}^\bullet O-N=$$

takes place with a third molecule of dimethylglyoxime, resulting in a new radical pair which is situated differently in the crystal and which has an e.s.r. spectrum differing from that of the original pair. Since the crystal structure of dimethylglyoxime is known, the e.s.r. spectra can be interpreted in detail and the nature of the radical pairs established. The first kinetic measurements on this transformation [234] covered the ranges 77–100 K for the H-compound and 190–220 K for the D-compound: the apparent activation energies were 6 kJ mol$^{-1}$ and 30 kJ mol$^{-1}$ respectively. It was inferred that $k^H/k^D$ was in the range $10^3$–$10^5$, though this was not verified directly by measurements at a common temperature. This behaviour was attributed to tunnelling and a tentative quantitative interpretation given. A more recent study [235] gives rates for the H-compound from 50 K to 145 K, together with an isolated value at 4.2 K. In the range 50–145 K the Arrhenius plot is strongly curved, the apparent activation energy increasing from about 1 kJ mol$^{-1}$ to 10 kJ mol$^{-1}$. These results could be fitted quantitatively by calculations based on an unsymmetrical Eckart barrier with $V_0 = 63$ kJ mol$^{-1}$, $\Delta H = -4$ kJ mol$^{-1}$, $b = 44$ pm, though it is not stated how the classical pre-exponential factor was estimated. It is also stated that the earlier results at 190–220 K for the deuteriated compound [234] are represented by the same model.

The rate predicted for the H-compound at 4.2 K is about one tenth of the observed rate, which is almost the same as that at 50 K. This discrepancy is

# Experimental evidence for tunnelling

probably due to the assumption of a continuous energy distribution at temperatures where effectively all systems must be in their lowest state: if this so between 4.2 K and 50 K the rate should be independent of temperature in this range. However, as pointed out by the authors [235], the calculated tunnel 'correction' at 4.2 K is $10^{764}$, so that a discrepancy of one power of ten cannot be regarded as very serious. In any event, the investigations of hydrogen-atom transfer described in this section form a coherent and convincing demonstration of the dominance of tunnelling processes at low temperatures.

There are a number of other investigations which, although less detailed than those quoted above, strongly suggest tunnelling by hydrogen atoms at low temperatures. A series of studies by Willard and his co-workers [236–239] deal with hydrogen abstraction from 3-methylpentane glasses by methyl radicals produced by irradiation of methyl iodide: the observations include the rate of decay of $CH_3^\cdot$, the growth of $C_6H_{13}^\cdot$, and the isotopic composition of the methane produced when $CD_3I$ or $C_6D_{14}$ was used. The apparent activation energies are given as $3 \pm 1$ kJ mol$^{-1}$ in the range 45–77 K, and $> 13$ kJ mol$^{-1}$ at 77–100 K. No deuterium abstraction from $C_6D_{14}$ could be detected at 77 K. In another investigation [240] deuterium atoms produced by the photolysis of DI were found to abstract hydrogen from $C_6H_{14}$ at 77 K, a process which should not be observable in the absence of tunnelling.

The abstraction of hydrogen from thin films of polyisobutene and polyepoxyethane by hydrogen atoms from the gas phase has been studied at 99 K, 123 K and 145 K, the concentrations of hydrogen atoms and of the radicals produced being monitored by observing their e.s.r. spectra [241]. The observed activation energies are 8 kJ mol$^{-1}$ and 5 kJ mol$^{-1}$ for the two polymers studied, both of which are much smaller than the values of 20–30 kJ mol$^{-1}$ commonly found for analogous reactions in the gas phase at 300–800 K. The authors give a quantitative interpretation in terms of tunnelling through parabolic or Eckart barriers, though this seems hardly justifiable in view of the paucity and low accuracy of the experimental results.

Some large isotope effects at rather higher temperatures have been observed in the oxidation of 4a-4b-dihydrophenanthrene by molecular oxygen in isooctane solution [242, 243]. Denoting the substance by $PH_2$ (or $PD_2$), the reactions involved in the presence of a high concentration of inhibitor SH are:

$$PH_2 + O_2 \rightarrow PH^\cdot + HO_2^\cdot$$
$$PH^\cdot + O_2 \rightarrow P + HO_2^\cdot$$
$$SH + HO_2^\cdot \rightarrow H_2O_2 + \text{inactive products}$$

with the first reaction (initiation) rate-determining. The isotope effects for the initiation reaction were found to be $k^H/k^D = 64$, 95 and about 250 at 263 K, 242 K and 221 K respectively: the maximum semi-classical values at these temperatures are 13, 17 and 22. Measurements for the hydrogen compound at eight temperatures between 191 K and 263 K showed marked deviations from the Arrhenius equation, the discrepancy at 191 K amounting to a factor of three. The results at the two higher temperatures yield $E_A^D - E_A^H = 11 \text{ kJ mol}^{-1}$, $A_A^D/A_A^H = 2.3*$, while for 242 K and 221 K the approximate values are $E_A^D - E_A^H \approx 20 \text{ kJ mol}^{-1}$, $A_A^D/A_A^H \approx 200$. The authors have given a detailed interpretation of their results in terms of the tunnel effect [244, 245], using the procedure developed by Johnston and his collaborators for simple gas reactions and described in the next section. They were able to reproduce the observed isotope effects and deviations from the Arrhenius equation, but their predicted value of $A_A^D/A_A^H$ is considerably too high, and it is doubtful whether the type of theoretical treatment used is suitable for a system of this complexity.

A study of the fate of the radicals $NH_2CONHO^{\cdot}$ produced by the action of X-rays on hydroxyurea (and its deuteriated analogue) [246] is reminiscent of the work on dimethylglyoxime described above, though reproducibility was poor and the range of temperatures studied (253–268 K) was much higher and narrower. The radicals originally produced occur partly as weakly interacting pairs 64 pm apart and partly as isolated radicals, which can be distinguished by differences in their e.s.r. spectra. In the course of time the spectrum of the isolated radicals increases in intensity, while that of the radical pairs decreases, and this is attributed to the transfer of a hydrogen atom from an 'undamaged' molecule to one member of a radical pair. The isotope effects reported are between 15 and 30, and the Arrhenius expressions given correspond to $E_A^D - E_A^H \approx 30 \text{ kJ mol}^{-1}$, $A_A^D/A_A^H \approx 6 \times 10^4$, which on the face of it suggest very large tunnel corrections. However, this interpretation is not favoured by the authors, who say 'the difference between the exponential factors is not regarded as significant', and who seem to imply in their discussion that tunnelling could not have appreciable experimental consequences in this type of process. In view of the very striking results obtained with irradiated dimethylglyoxime further work on hydroxyurea would be of interest.

## 5.3 Hydrogen atom transfers in simple gas reactions

There is available a large amount of experimental data on the kinetics of gas reactions of the type

$$XL + Y \rightarrow X + LY$$

of which the simplest relate to the reaction $H + H_2$ and its isotopic variants. Although this would appear to be an obvious field for investigating the part played by the tunnel effect, the present position is somewhat confused for a number of reasons. It is rarely possible to study homogeneous gas reactions at temperatures below about 300 K, since they either become too slow, or are complicated by surface reactions. They therefore do not reveal the striking isotope effects or deviations from the Arrhenius equation which have been observed for low-temperature processes in condensed phases, as discussed in

---

* In considering the results for these two temperatures the authors state that $A_A^D/A_A^H = 1.6$, but a value of 2.3 agrees better with experiment.

# Experimental evidence for tunnelling

Section 5.2. On the other hand, the free relative motion of two reactants in the gas phase justifies the application of tunnel corrections based on a continuous distribution of energies, and the simple nature of the systems involved makes it feasible to construct an energy surface, either *a priori* from quantum theory or by some empirical or semi-empirical procedure.

It is thus possible to envisage for simple gas reactions a programme which is much more ambitious than anything suggested so far in this book. Ideally this programme runs as follows. An energy surface is first constructed for all configurations of the system X——L——Y, and is analysed to establish the configuration of the transition state, together with its energy relative to the reactants, its moments of inertia and the frequencies of its normal vibrations. This information, together with the molecular parameters of the reactants, is sufficient to make an absolute prediction of the reaction velocity from the semi-classical expression of transition state theory, Equation 3.17. This procedure can be carried out for different isotopic masses, or if only the isotope effect is required, expressions such as Equation 4.11 or Equation 4.15, containing only the vibration frequencies, can be used. Comparison of the results with experiment should indicate whether an appreciable tunnel correction is required, and in fact this correction can be predicted from the energy surface if it is assumed that a treatment of tunnelling in one dimension is adequate. If the tunnel correction is not too large, so that the relevant part of the barrier approximates to a parabola, the correction is simply applied by including an 'imaginary frequency' among the factors in the vibrational partition function, as may be seen by comparing Equations 3.23 and 3.28. If, as is often the case, the tunnel correction turns out to be considerable, the energy surface can be used to construct the energy profile along the path of lowest energy. The profile can then be 'straightened out' into a one-dimensional barrier and the tunnel correction computed numerically, if necessary by fitting to a realistic analytical shape such as the Eckart barrier.

Unfortunately this programme represents an ideal rather than a real situation. With the possible exception of recent calculations for the system of three hydrogen atoms, *a priori* calculations of energy surfaces by quantum mechanics are not sufficiently accurate to yield reliable estimates of the real and imaginary frequencies of the transition state, and this applies with equal force to the profile along the whole of the reaction path. (This is not surprising, since the desired accuracy represents something like 0.1% of the total electronic energy, even in the most favourable cases.) It has therefore been common practice to adjust the theoretical energy surface so as to fit the observed activation energy, rate, or isotope effect at a particular temperature. Since the *a priori* calculations of an energy surface are at best very laborious, semi-empirical or empirical surfaces have frequently been used, and these contain one or more adjustable parameters which have to be chosen so as to agree with some aspect of the experimental data.

Difficulties can also arise in applying the tunnel correction. The energy profile provided by the theoretical or semi-empirical surface sometimes leads to one-dimensional tunnel corrections which are too great to be consistent with experiment, and some authors have arbitrarily modified this part of the surface to give better agreement. Moreover, there is no theoretical justification for straightening out the reaction path and treating the barrier thus obtained as a separate one-dimensional problem, except when the tunnel correction is small. There is no simple method for circumventing this difficulty, but some suggested methods for doing so are mentioned in the next two sections and are discussed in more detail in Chapter 7.

The programme outlined above assumes, like all the discussion so far in this book, that the reaction velocity can be calculated by multiplying the expression given by transition state theory by a suitably computed tunnel correction. Many modern treatments of reaction kinetics do not employ transition state theory, but proceed in a more fundamental way by considering individual trajectories of the reacting species which can lead to reaction and computing the reaction velocity as a weighted mean over all possible trajectories.* In one form of treatment the trajectories are calculated by classical mechanics, and a tunnel correction can in principle be applied to each trajectory. In the most fundamental form of the theory the problem is treated quantum-mechanically throughout: in this event the tunnel effect does not appear as a separate issue, since it is automatically included in the computation of the transition probabilities. So far these detailed considerations of molecular dynamics have found little application outside the simplest possible case of $H + H_2$. Under some conditions they produce results differing little from those of transition state theory, but there can be some important differences, especially at low temperatures. Some of these problems will be discussed in Chapter 7, but in the present chapter we shall adhere to the general theme of this book, namely the application of corrections for tunnelling to the conventional expressions of transition state theory.

The literature in this field is now vast, and no attempt has been made to cover it completely. Section 5.3.1 deals with the reaction $H + H_2$ and its isotopic variants, and Section 5.3.2 with a few other reactions involving hydrogen atoms or the simple radicals $CH_3$ and $CF_3$. For the reasons already stated the quantitative conclusions arrived at have varied with time and from one investigator to another. Nevertheless, the general picture is now fairly clear, and there can be no doubt that tunnel corrections of appreciable magnitude are necessary if transition state theory is to be applied to simple gas reactions involving transfer of hydrogen atoms.

* In addition to the overall rates these theories yield much more detailed information, for example the reaction probabilities of individual vibrational and rotational states and the angular distribution and quantum states of the products. Some of this information can be compared with the results of molecular beam experiments.

### 5.3.1 The reaction $H + H_2$ and its isotopic variants

This reaction has been a touchstone for theories of chemical kinetics ever since London [247] showed in 1929 that the application of quantum theory to a system of three hydrogen atoms predicted an activation energy much less than the dissociation energy of the hydrogen molecule, in agreement with experiment. Recent reviews dealing wholly or largely with this reaction have appeared in 1966 [248], 1967 [249], 1970 [250], 1976 [251] and 1977 [252]. The following four processes have been studied both experimentally and theoretically:

1. $\qquad H + H_2 \rightleftharpoons H_2 + H$
2. $\qquad D + D_2 \rightleftharpoons D_2 + D$
3. $\qquad D + H_2 \rightleftharpoons DH + H$
4. $\qquad H + D_2 \rightleftharpoons HD + D$

Reactions 1 and 2 have of course the same velocity constants in both directions, while for Reactions 3 and 4 the forward and reverse constants are related by equilibrium constants which can be accurately predicted. The first two reactions involve no overall change, but can be followed by observing the interconversion of the *ortho-* and *para-*forms, first studied kinetically in 1930–35 [253–256]. Reactions 3 and 4 are followed by some form of isotopic analysis. In more recent work [257–264] low-temperature gas chromatography has been used both for isotopic analysis and for analysing *ortho–para* mixtures: electron spin resonance has also been used to detect atoms. Measurements have been carried out under two sets of conditions. In the first of these the H- or D-atoms present are those in thermal equilibrium with the species $H_2$, $D_2$ or HD, and since these atom concentrations are low this method can only be used at high temperatures, in practice 700 K to 1000 K. It has been applied to the *ortho–para* interconversion of $H_2$ and $D_2$ and also to the rate of the reaction

$$H_2 + D_2 \rightleftharpoons 2HD$$

The latter yields a combination of the forward and reverse rate constants for Reactions 3 and 4. In the second type of investigation H- or D-atoms are produced artificially (for example on a heated wire or in a discharge) and are injected into a stream of gas passing along a flow tube, the hydrogen atom concentration being monitored by electron spin resonance or by a thermal gauge. Since much larger concentrations of atoms can be used measurements can be extended to lower temperatures, and a range of 167–745 K has been covered. Both types of method offer some difficulties of technique and interpretation, but fairly reliable velocity constants are now available for Reactions 1–4 over a wide temperature range.

Attempts to predict or interpret this experimental material have been numerous and confusing: a recent review [251] gives more than 400 references. As early as 1933 it was pointed out [40] that the kinetic information then available about the $o$-$H_2$–$p$-$H_2$ interconversion suggested considerable deviations from the Arrhenius relation which could be explained in terms of the tunnel effect. This suggestion was based on inadequate evidence, but the more complete results now available do show the expected behaviour, in that the apparent activation energies vary with temperature by some 20–30%. Apart from this, it is not easy to apply any of the simple tunnelling criteria listed in Sections 3.7 and 4.3.5. This is because the reaction $H + H_2$ is peculiar in two respects. In the first place one of the reactants is an atom, so that the formation of the transition state H——H——H involves a change from one to three real vibrational degrees of freedom, all of which have frequencies which are isotopically sensitive. Secondly, because all three atoms have comparable masses, the effect of isotopic substitution on the frequencies cannot be represented even approximately by a factor of $2^{1/2}$, except when all three hydrogen atoms are replaced by deuterium. The problem is thus more complicated than for the transfer of a proton or hydrogen atom between two heavy centres, and progress cannot be made without deriving or assuming an energy surface for the system.

Early computations used empirical or semi-empirical energy surfaces which are now known to be inaccurate. They will therefore not be described here, though it may be noted that tunnel corrections of varying magnitudes were always found to be necessary in the lower part of the temperature range. An account of the position up to about 1965 is given by Johnston [248]. The first *ab initio* calculation with any pretensions to accuracy was that of Shavitt and his collaborators [265]. This still predicts high-temperature activation energies which are about 11% too high, and it has been common practice to modify this surface arbitrarily by multiplying the potential energies along the reaction path by 0.89: the result is often referred to as the 'scaled Shavitt surface'. Shavitt [266] has applied this surface to calculate by transition state theory the rates of all the isotopic variants of the reaction $H + H_2$, including those involving tritium, which are not at present accessible to experiment. The frequencies which he calculates for the reactants and transition states of Reactions 1–4 are given in Table 5.1. These values can be used to calculate the isotopic rate ratios for the various reactions from the semi-classical transition state expression. Since the bending vibration of the transition state is doubly degenerate and must be counted twice over there is not a large loss of zero-point energy in going from the initial to the transition state, in spite of the high vibration frequencies of $H_2$ and $D_2$: in fact, for the reaction $H + D_2$ the transition state actually has slightly more zero-point energy than the reactants. For this reason the isotope effects are not expected to be very large. For example, if we compare the reactions $H + H_2$ and $H + D_2$ the calculated

Table 5.1. *Frequencies (cm$^{-1}$) calculated from the scaled Shavitt surface*

| Species | Symmetrical stretch | Bend | Reaction co-ordinate |
|---|---|---|---|
| H–H | 4370 | — | — |
| D–D | 3091 | — | — |
| H—H—H | 2012 | 965 | 1452i |
| H—H—D | 1732 | 924 | 1377i |
| H—D—D | 1730 | 737 | 1090i |
| D—D—D | 1423 | 683 | 1027i |

values of $k_1/k_4$ are 7.6 at 280 K, 4.9 at 400 K, and 3.0 at 700 K. At the highest temperatures these agree fairly well with experiment, but there is a discrepancy which increases with decreasing temperature: thus the observed values of $k_1/k_4$ are about 6.5 at 400 K and 11.5 at 280 K. These discrepancies, which also appear when the reactions $D + H_2$ and $D + D_2$ are considered, are strongly indicative of considerable tunnel corrections, since the transition state treatment should give more reliable values for isotopic rate ratios than for absolute rate constants.

However, there are difficulties in producing a more quantitative treatment of these findings. The values of the imaginary frequency $iv_\ddagger$ in Table 5.1 correspond to values of $u_\ddagger = hv_\ddagger/kT$ in the range 4–8 in the lower part of the temperature range, which on the basis of a parabolic barrier corresponds to rather large tunnel corrections (cf. Section 3.6.1 and Table 3.1). The application of these corrections predicts rates and isotope effects which are much too large: this is only to be expected when the tunnel corrections are considerable, since the energy profile along the reaction path can only be approximated by a parabola over a small portion near its apex. A more realistic reaction profile is provided by the Eckart barrier (Equation 2.46), but if this is chosen so as to have the same height and curvature at the top as the scaled Shavitt surface the tunnel corrections obtained are again too large, though the discrepancy is smaller than for the parabolic barrier. Shavitt [265] therefore calculated tunnel corrections on the basis of an Eckart function which gave a reasonable fit over the top half of the barrier, but which levelled off some 20 kJ mol$^{-1}$ above the true floor: this procedure obviously underestimates tunnelling through the base of the true barrier, though this will be unimportant except at quite low temperatures. This rather arbitrary choice of barrier gave tunnel corrections of considerable magnitude at the lower temperatures and achieved reasonably good agreement with experiment both for the absolute velocity constants and for the isotopic rate ratios.

A more satisfactory method of calculating the tunnel correction has been applied to the same surface by Quickert and Le Roy [267, 268]. They employed a numerical method [52–54] to calculate the permeabilities and

tunnel corrections for the actual reaction profile of the scaled Shavitt surface*
and obtained excellent agreement with the observed isotope effects. Examples
of their calculated tunnel corrections for Reactions 1–4 are 7.21, 2.99, 5.52 and
3.54 at 280 K; 2.72, 1.76, 2.41 and 1.90 at 400 K; and 1.35, 1.19, 1.31 and 1.21 at
800 K.

More recent work by Liu [269] has led to an *ab initio* surface which is more accurate
than that of Shavitt *et al.*, its accuracy being estimated at about 1 kJ mol$^{-1}$. The
original calculations referred only to linear configurations of the three atoms, but have
been extended empirically to non-linear systems [270], as is necessary to obtain the
bending frequencies of the transition state. Transition state calculations have been
carried out for this surface [271], and comparison with the observed rate constants for
$D + H_2$ and $H + D_2$ definitely establishes the need for tunnel corrections. These are
considerably smaller than those calculated by Quickert and Le Roy [267, 268]; for
example for $D + H_2$ at 325 K $Q_t$ is about 1.8 instead of 3.6. One-dimensional tunnel
corrections for the reaction profile of Liu's surface do not appear to have been
computed, probably because of growing doubts as to the quantitative validity of
combining transition state calculations with a one-dimensional tunnel correction for a
'straightened-out' reaction path.

It is probable, however, that the quantitative aspects of these calculations
must not be taken too seriously. The difficulties which arise will be considered
in more detail in Chapter 7, and only the main points considered here. As first
pointed out clearly by Johnston and Rapp [89], the use of a one-dimensional
tunnel correction for the 'straightened-out' reaction path is not really
legitimate. This is because the equation for motion along the reaction co-
ordinate is mathematically separable from the general equations of motion
only when the potential energy is a quadratic function of all the co-ordinates,
including the reaction co-ordinate. This condition is satisfied for separated
reactants or products, when the potential energy is independent of the reaction
co-ordinate, and also for a small region in the neighbourhood of the transition
state in which the energy barrier can be approximated by a parabola, but it is
not satisfied for the remainder of the reaction path. Examination of the
theoretical energy surfaces for the reaction $H + H_2$ shows that the parabolic
approximation is fairly good only for about 15 pm on either side of the saddle
point. The dimensions of the region over which tunnelling is significant are
given approximately by the de Broglie wavelength of the hydrogen atom,
which is 100 pm at 300 K and 55 pm at 1000 K, so that the usual procedure
appears to be unjustifiable. Johnston and Rapp [89] proposed an empirical

---

\* When tunnelling over the whole length of the reaction path is considered the effective particle
mass in real space varies along this path: for example, for the reaction $H + H_2$ it is $2m_H/3$ at the
ends of the path and $m_H/3$ in the centre. Quickert and Le Roy preferred the equivalent procedure
of taking a reduced mass which is independent of the reaction co-ordinate and scaling the reaction
co-ordinate by a factor which varies along its length. For this reason the barrier profiles which
they use depend on the isotopic masses and are not symmetrical for the unsymmetrical systems
$H + D_2$ and $D + H_2$.

# Experimental evidence for tunnelling

procedure in which the tunnel correction is averaged over a number of cuts through the energy surface parallel to the reaction co-ordinate, as indicated by the broken straight lines in Fig. 5.2, but this method lacks any clear justification. A very recent suggestion [272] is that the tunnel trajectory should be taken not along the path of lowest potential energy, but along the limit of classical vibration amplitudes for a given vibrational state, as indicated by the broken curve in Fig. 5.2. This makes the effective barrier higher and narrower and is stated to give better agreement with the recent quantum calculations of Schatz and Kuppermann [273]. This suggestion has some physical plausibility, but it is difficult to see how it avoids the problem of non-separability over an extended region.

Much of the recent theoretical work on $H_3$-kinetics has been devoted to methods which should in principle be more rigorous than transition state theory (with or without tunnel corrections) and which are concerned with the

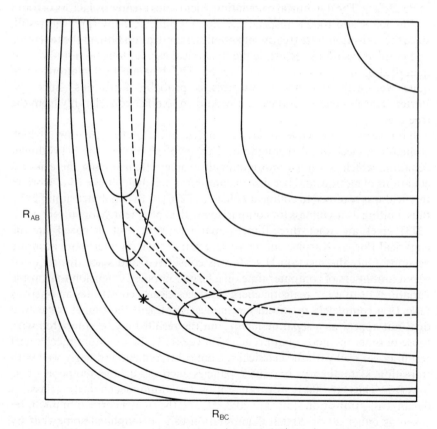

Fig. 5.2  Alternative tunnelling paths for the reaction $AB + C \rightarrow A + BC$

detailed behaviour of individual collisions as well as with the averaged result expressed by the velocity constant. A review of developments in this field has been given by Connor [274]. The emphasis has frequently been on a comparison of the results of different treatments rather than on comparison with experiment, and the energy surfaces have therefore sometimes been chosen for ease of computation rather than for their approximation to reality. It is difficult to gauge the relevance of these developments to the general theme of tunnelling in chemical reactions, and only a few examples will be quoted here.

In one type of dynamical calculation [87] the initial vibrational and rotational states of the $H_2$ molecule were taken to be those given by quantum theory, but the individual trajectories and collisions were treated classically, so that no tunnel corrections are involved. The calculations used a semi-empirical energy surface devised by Porter and Karplus [275] and the results were compared with transition state theory (without tunnel corrections) for the same surface. The dynamical calculations yield rates greater by factors of 6 and 1.25 at 300 K and 1000 K respectively, but it is difficult to interpret this result, since the transition state treatment assumes quantization both in the reactants and in the transition state, while the dynamical calculations do not allow for quantization in the course of the reaction. The dynamical calculations agree quite well with experiment, but this is probably fortuitous, since the Porter–Karplus energy surface is now known to differ considerably from the true one.

It is possible in principle to carry out a dynamic trajectory treatment which is quantum-mechanical throughout. This involves solving the Schrödinger equation, which is a much more difficult problem than solving the classical equations of motion, and because of computational difficulties earlier work in this field was confined to collinear collisions [276] or to two dimensions [277], thus limiting its usefulness for comparison with experiment. More recent work [273] gives an exact three-dimensional treatment, but is based on the empirical Porter–Karplus energy surface rather than on later more accurate versions. Only the reaction $H + H_2$ was considered, and since this required about four hours of computer time on a large computer, extension to isotopic variations or to more realisitic energy surfaces clearly constitutes an onerous task. In a completely quantum-mechanical treatment the tunnel correction does not appear as a separate factor, but the results may be compared with those of other methods of approach. Above 600 K the quantum-mechanical result agrees well with the classical dynamical treatment and fairly well with transition state theory, but discrepancies increase as the temperature is lowered. Thus at 300 K the rate ratios (quantum-mechanical):(classical dynamical):(transition state) are 20:3.3:1, and the tunnel correction might be taken as either 20 or 20/3.3. Computation can be simplified somewhat by

# Experimental evidence for tunnelling

treating the trajectories by the so-called semi-classical* version of quantum theory, reviewed recently by Miller [278]. This employs asymptotic solutions of the Schrödinger equation which become exact only when the de Broglie wavelength is much less than molecular dimensions, a condition which is certainly not satisfied for the $H_3$-reaction. However, for a given energy surface this procedure appears to give results quite close to those of a strict quantum treatment [279]. The position resembles that for the BWK approximation to the permeability of one-dimensional barriers (Section 2.5), which rests on essentially the same assumptions: for realistic barriers Equation 2.99 gives values very close to the exact solution even when the parameters of the system do not appear to satisfy the necessary conditions.

An alternative approach, which avoids the need to consider individual trajectories, is to seek a modification of conventional transition state theory which will include non-classical behaviour along the reaction path without running into difficulties about the non-separability of the reaction co-ordinate. A general formulation of such a quantum-mechanical version of transition state theory has been given by Miller [280], but in order to obtain useful results it is necessary to introduce semi-classical approximations [281]. The last approach has been applied to the $H_3$-reaction [90], and a note added in proof states that a full three-dimensional calculation for the reaction $D + H_2$ with the most accurate energy surfaces available [269, 270] gives exact agreement with experiment at 200 K and 1000 K. However, details or further calculations by this method do not appear to have been published. The derivation of expressions for quantum-mechanical transition state theory seems to owe a good deal to intuition, though an alternative formulation [282] offers some analogies to the kind of tunnelling paths suggested by Johnston and Rapp [89] and by Marcus and Coltrin [272].

Our theoretical understanding of the kinetics of $H + H_2$ and similar reactions is thus at present in a state of flux, with several rival theories in the field. The relation between these theories is not always clear, and in this context the terms 'tunnel effect' or 'tunnel correction' admit of several interpretations. However, if we define the tunnel correction as the ratio between the observed rate or isotope effect and the value predicted by conventional transition state theory, then there is no doubt that these

---

* The terms 'semi-classical' and 'quasi-classical' have been used in a confusing way by different authors. In addition to the sense in which it is used here the description 'semi-classical' is often applied to dynamical calculations in which the initial state of the system is treated by quantum theory, while the calculation of the trajectories is carried out by classical mechanics. In the present book 'semi-classical' has also been used to describe the usual version of transition state theory, in which quantum theory is applied to the real vibrations of the reactants and the transition state, but not to motion along the reaction co-ordinate (i.e. the tunnel correction is ignored.) *Caveat lector*!

corrections are considerable at the lower end of the experimental temperature range, say 200–400 K, though it is difficult to assign definite values to them.

### 5.3.2 Other gas reactions involving hydrogen abstraction

Apart from reactions involving three hydrogen atoms there has been much experimental and theoretical work on hydrogen atom transfers in other simple gas reactions, only a selection of which will be quoted here. Since heavier atoms are now involved the picture of the process is a simpler one, but the possibility of any *ab initio* calculation of energy surfaces is remote, so that the theoretical interpretation has to be based on surfaces which are at best semi-empirical.

The simplest reaction in this category is the reaction of photochemically produced chlorine atoms with the molecules $H_2$, HD, HT, $D_2$, DT, and $T_2$. Since the chlorine atom can abstract either atom from the unsymmetrical species there are in principle nine independent rate constants for the initial step

$$L_1L_2 + Cl \rightarrow L_1 + L_2Cl,$$

where $L_1$ and $L_2$ can each be H, D, or T. However, this step is followed by a fast reaction

$$L + Cl_2 \rightarrow LCl + Cl$$

so that the overall reaction is

$$L_1L_2 + Cl_2 \rightarrow L_1Cl + L_2Cl$$

independent of whether $L_1$ or $L_2$ is abstracted in the slow step. For this reason only five independent isotope effects can be measured by the usual procedures. In practice the isotopic rate ratios $H_2/HD$, $HD/D_2$, $H_2/HT$, $DT/D_2$ and $T_2/D_2$ were obtained by irradiating the corresponding mixtures and carrying out isotopic analyses of the unreacted hydrogen. The first measurements [283–285] employed only hydrogen and deuterium, but subsequently [286] all three isotopes were used. The results refer to two to five temperatures in the range 237–342 K, but are not sufficiently extensive to detect any deviations from the Arrhenius equation. Later [287] the intramolecular isotope effect for the reactions

$$DH + Cl \rightarrow D + HCl$$

and

$$HD + Cl \rightarrow H + DCl$$

was determined by measuring the HCl/DCl ratio in flow systems in the presence of nitrogen dioxide, which scavenges the hydrogen atoms and prevents the subsequent reactions $L + Cl_2 \rightarrow LCl + Cl$.

The transition states in these reactions can be characterized by four force constants ($f_1$, $f_2$ and $f_{12}$ in Equation 3.14, together with a bending force constant), and since six observed isotope effects are available it is possible to choose the four force constants empirically so as to give the best agreement with experiment. This was carried out by an iterative procedure [288], in which the tunnel correction was calculated by fitting a one-dimensional Eckart barrier to the imaginary frequency.* It was concluded that only moderate tunnel corrections were involved, the maximum value of $Q_t$ being 2.2 for $Cl + H_2$ at 250 K. Attempts were also made to interpret the experimental results in terms of two types of semi-empirical energy surface. The first, due to Sato [289] is a modification of the original London treatment [247]. The second is the so-called bond-energy-bond-order (BEBO) method, in which the properties of the system along the reaction path are calculated in terms of the orders of the partial bonds in H——H——Cl, assumed to add up to unity: this treatment had previously been applied to the same reaction by Johnston and Parr [290]. Although both of these surfaces were adjusted so as to reproduce the observed activation energy [291], neither was found to be fully consistent with all six observed isotope effects, which casts some doubt on their use in these systems, in particular for drawing conclusions about tunnel corrections. If no tunnel correction was included the isotope effects predicted from the Sato surface were considerably too low, for example 7.1 and 18.3 for $H_2/D_2$ and $H_2/T_2$ at 298 K, compared with the observed values of 9.6 and 26.4. However, if tunnel corrections were calculated from a one-dimensional unsymmetrical Eckart barrier fitted to the curvature of the barrier at the top the predicted isotope effects are too high, i.e. the tunnel correction is overestimated. Better agreement was obtained by following the procedure of Johnston and Rapp [89], in which the tunnel correction is averaged over a number of cuts parallel to the reaction co-ordinate (cf. Fig. 5.2) though, as mentioned in connection with the $H + H_2$ reaction, there is no strict justification for this procedure.

Attempts have been made to apply classical trajectory calculations and quantum-mechanical methods to the reactions of chlorine atoms with hydrogen [292–294]. Since these are all restricted to collinear configurations and are based on empirical energy surfaces their relevance to experiment is doubtful. For a given Sato surface a comparison was made between the rates predicted by quantum mechanics, by classical trajectories, and by transition state theory: no tunnel corrections were made in the last two and all refer to collinear systems. For the reaction $Cl + H_2$ in the range 200–450 K the

---

* These calculations supersede the earlier ones [286] in which only five observed isotope effects were used and the empirical force constants were determined by trial and error. This led to very flat barriers and hence to very small tunnel corrections.

following ratios were found:
(quantum mechanical):(classical) = 3.57 to 1.81
(quantum mechanical):(transition state) = 4.79 to 1.81.

These figures are consistent with the tunnel corrections deduced by fitting the force constants of the transition state to the observed isotope effects [288].

Somewhat less extensive information is available for the reaction of fluorine atoms with $H_2$, $D_2$, HD, and DH [295, 296] and similar theoretical treatments have been given [297, 298]. The conclusions are broadly similar to those for the corresponding reactions of chlorine atoms.

Many other gas reactions have been studied in which a hydrogen atom is abstracted from a molecule by an atom or radical. In the present context most interest attaches to isotope effects, a few of which will be mentioned here. There are a few instances in which the raw results give good evidence for an appreciable tunnel correction in terms of the criteria of Section 4.3.5. Thus in the reaction of $CF_3$ with $CH_2D_2$ or $CHD_3$ [299, 129] the isotope effect for the abstraction of H or D can be measured by isotopic analysis of the products, giving the following results:

| $T/(K)$ | 315 | 325 | 328 | 368 | 380 | 382 |
|---|---|---|---|---|---|---|
| $k^H/k^D$ | 16.3 | 18.5 | 15.7 | 11.8 | 10.5 | 10.3 |

These values are approximately twice as great as the maximum values predicted on a semi-classical basis (cf. Table 4.2), and they give $E_A^D - E_A^H = 8.0$ kJ mol$^{-1}$, $A_A^D/A_A^H = 1.5$, suggesting a moderate tunnel correction. Similarly, a comparison of the reactions H + $CH_4$ and D + $CH_4$ [300] over the range 500–732 K gave $A_A^D/A_A^H = 2.5 \pm 0.7$.* Further general evidence comes from analyses [301, 302] of kinetic results over a wide temperature range for the reactions H + $CH_4$, H + $C_2H_6$, $CH_3$ + $C_2H_6$ and $CH_3$ + neopentane. These reveal deviations from the Arrhenius equation which can be attributed to tunnelling, though such evidence is not compelling unless supported by isotope effects.

Most attempts to derive tunnel corrections for these hydrogen abstraction reactions have depended on interpreting observed isotope effects in terms of empirical energy surfaces, and their conclusions must be qualified by our

---

* The value of $A_A^D/A_A^H$ given here comes from the least-squares treatment of the experimental results given in the original paper [300]. The authors, however, have arbitrarily corrected the value of $A_A^D$ from $1.6 \times 10^{11}$ to $4.5 \times 10^{10}$ dm$^3$mol$^{-1}$s$^{-1}$ so as to give $A_A^D/A_A^H = 2^{-1/2}$, as predicted by classical collision theory. Their least-squares treatment leads to closely similar activation energies for the two isotopes, with $E_A^H$ the greater by about 2 kJ mol$^{-1}$. In these reactions there are no isotopically sensitive vibrations in the initial state, so that because of the presence of such vibrations in the transition state a semi-classical treatment would predict $E_A^H > E_A^D$ by at least 10 kJ mol$^{-1}$. The small difference actually observed is consistent with appreciable tunnel corrections, which would lower $E_A^H$ more than $E_A^D$.

# Experimental evidence for tunnelling

ignorance of how closely these surfaces represent the real situation. A review of some of these calculations is given by Johnston [248]. Two types of surface have been commonly used. The first is Sato's modification [289] of the original London treatment, adjusted so as to give the observed activation energy: this is commonly referred to as LEPS (London–Eyring–Polanyi–Sato). The second is based on the bond-energy-bond-order (BEBO) method, which uses the spectroscopic properties of the reactants and products together with various assumptions about the relations between bond order and bond energy [290]. (The BEBO method does not yield a complete energy surface, but only the profile along the reaction path, together with the bending force constant in the transition state.) In most instances a transition state treatment based on either of these surfaces, but omitting any tunnel correction, seriously underestimates the isotope effect at moderate temperatures. For example, in the reactions of $CF_3$ with methane [129] the discrepancy amounts to a factor of 3 at 300 K and 50% at 500 K. Similar results were found for the reaction of $CD_3$ with $C_2H_6$ and $C_2D_6$ [303]. In these and many other reactions much closer agreement can be reached by including tunnel corrections, but there is no general agreement as to whether these should be calculated by fitting a one-dimensional barrier to the reaction path, or by the method of parallel cuts suggested by Johnston and Rapp [89]. Moreover, in more recent calculations on some reactions of alkyl and hydroxyl radicals [304, 305] it is claimed that BEBO or LEPS treatments agree better with experiment if tunnel corrections are omitted, although the values obtained for the imaginary frequencies of the transition state suggest that such corrections should be of considerable magnitude. The present position is therefore somewhat obscure.

## 5.4 A general survey of hydrogen isotope effects in solution

This chapter has dealt so far with systems in which striking consequences of tunnelling have been observed, especially at low temperatures, and with a few simple gas reactions which have been thoroughly studied experimentally and for which a detailed theoretical interpretation has been attempted. There also exists a large body of evidence for tunnelling relating to reactions in solution which have not been studied so extensively, and which are too complicated (either intrinsically or by the involvement of solvent molecules) for detailed theoretical treatment. Most of this evidence relates to hydrogen isotope effects, and involves the observation of unexpectedly large values of $k^H/k^D$, $E_A^D - E_A^H$ or $A_A^D/A_A^H$, or of the corresponding quantities for tritium. In a few instances deviations from the Arrhenius equation have also been detected. A review of such evidence was given in 1974 [117], and several more examples have come to light since that date.

The material in Tables 5.2, 5.3, and 5.4 has been classified as proton transfers, hydrogen atom transfers and oxidation reactions. When proton transfers (acid–base reactions) are thermodynamically favourable they can be observed directly, for example by spectrophotometric or conductimetric methods. Frequently, however, equilibrium lies very much in favour of the reactants, and in this event less direct methods must be used. One of these consists of the use of a scavenger to react rapidly with one of the products as soon as it is formed. A common example is the use of iodine or bromine to react with the carbanion produced from a weak carbon acid: the rate of halogenation is frequently independent of halogen concentration, and can then be equated to the rate of proton transfer. If the proton is lost from a chiral centre its return will be accompanied by a change in optical rotation, so that observations of racemization or mutarotation can be used to study the kinetics of proton-transfer processes. If the solvent contains exchangeable hydrogen and more than one isotope of hydrogen is present, the isotopic composition of both solute and solvent will change with time if proton-transfer processes are taking place. The commonest example of this is the use of rates of detritiation of organic solutes to measure the rate of triton transfer from the solute to the solvent or to other bases present in solution. Finally, there are some reactions in which proton transfer is accompanied by other changes which are readily observable, as in base-catalysed $\beta$-eliminations of the type

$$\diagup\!\!\!\!\!\diagdown\!\mathrm{CL} - \mathrm{C}\mathrm{X}\diagup\!\!\!\!\!\diagdown + \mathrm{B} \to \diagup\!\!\!\!\!\diagdown\mathrm{C} = \mathrm{C}\diagup\!\!\!\!\!\diagdown + \mathrm{BL}^+ + \mathrm{X}^-.$$

The proton-transfer reactions in Tables 5.2, 5.3 and 5.4 include examples of all these procedures.

The study of hydrogen atom transfers in solution often involves the reaction of atoms or radicals produced by radiation. Isotope effects have commonly been determined by competition methods in which the atoms or radicals react with a mixture of hydrogen and deuterium or tritium compounds, and isotopic analysis is carried out on the product or the reactant. Hydrogen atoms are also frequently transferred in the individual steps of chain reactions, notably polymerizations, and when inhibitors are present their reactions also commonly involve hydrogen atom transfer. As will be seen in the next paragraph, care is necessary in interpreting isotope effects for this class of reaction.

Oxidation reactions have been classified separately, since even when the presence of a considerable hydrogen isotope effect points to the transfer of hydrogen in some form, it is not always clear whether it is $H^+$, $H$ or $H^-$ which is transferred [178]. This ambiguity is not important for a general consideration of tunnel corrections.

It is believed that all the kinetic information in Tables 5.2, 5.3, and 5.4 relates to single-stage processes. It is, however, of interest to investigate how far any conclusions about tunnel corrections would be vitiated if the reaction being considered actually took place in more than one stage. This problem has been investigated particularly for proton-transfer reactions. A common example of a two-stage process is electrophilic aromatic substitution, in which proton transfer is preceded by a reversible step which does not exhibit an appreciable isotope effect. This example was considered in Section 4.1, and Equation 4.4 shows that the observed isotope effect will aways be smaller than the value for the proton transfer itself. Similarly, it is easy to show that the observed value of $E_A^D - E_A^H$ will be smaller than the 'true' value, and the observed value of $A_A^D/A_A^H$ closer to unity. All these effects are in the opposite direction to those caused by tunnelling. The same results emerge if a reversible proton transfer is followed by a reaction with no isotope effect. A common example is provided by the base-catalysed halogenation of ketones and similar substances in which a carbanion is formed by proton abstraction, followed by reaction of the carbanion with halogen. At sufficiently high halogen concentrations the initial proton transfer is rate-determining and its isotope effect can be measured directly, but at low halogen concentrations the halogenation becomes at least partly rate-determining and the observed isotope effect is decreased. In all of these instances the 'true' isotope effect can in principle be determined if the reaction is studied over a sufficiently wide range of conditions [122, 159], though this is less simple when the proton transfer constitutes the first step because of the possibility of isotopic exchange between the substrate and the solvent [156].

There are, however, at least two cases in which the presence of more than one stage can give rise to an abnormally high isotope effect and hence can simulate an appreciable tunnel correction. In the first of these the reactants form a transient intermediate by a reaction having a normal isotope effect, and this intermediate is then partitioned between two final products by reactions having different isotope effects. If the formation of only one of the final products is being monitored it is possible to observe an abnormally high isotope effect even when each individual stage exhibits a normal isotope effect. The simplest reaction scheme of this kind is as follows:

$$A \xrightarrow{k_1} B \begin{array}{c} \xrightarrow{k_2} C \\ \xrightarrow{k_3} D \end{array}$$

in which $k_2 + k_3 \gg k_1$, so that the intermediate B is never present in appreciable concentrations. If we write $k_1^H/k_1^D = \alpha_1$, $k_2^H/k_2^D = \alpha_2$, $k_3^H/k_3^D = \alpha_3$ and $k_2^H/k_3^H = \beta$, the usual steady state treatment gives the following expression for the isotope effect on the formation of C.

$$\frac{k_c^H}{k_c^D} = \frac{k_1^H k_2^H (k_2^D + k_3^D)}{k_1^D k_2^D (k_2^H + k_3^H)} = \frac{\alpha_1(\beta + \alpha_2/\alpha_3)}{\beta + 1}. \quad (5.3)$$

If $\alpha_2/\alpha_3 > 1$ the observed isotope effect will be greater than $\alpha_1$, and the maximum effect is observed if $\alpha_3$ has its minimum value of unity and $\beta \ll 1$. Under these last conditions the observed isotope effect will be just $\alpha_1 \alpha_2$, i.e. the product of two normal isotope effects. $E_A^D - E_A^H$ will also be abnormally large, since it is the sum of two normal values, but $A_A^D/A_A^H$ will not be markedly abnormal, being the product of two quantities each close to unity.

An example of this situation has been recently found by Ahlberg and his collaborators [306–310] in the reactions of 1-(2-acetoxy-2-propyl)-indene and similar substances with bases. Proton transfer initially forms a transient ion pair, which can then react further either by 1, 2-elimination of acetic acid, or by 1,3-proton migration to give an isomer of the original substance. The formation of the second product was found to have isotope effects of

Table 5.2. *Key to reactions in Tables 5.3 and 5.4 ($L = H$, $D$ or $T$)*

| Reaction | | References |
|---|---|---|
| *Proton transfers* | | |
| I | $\overline{(CH_2)_3 COCLCO_2 Et} + CH_2 ClCO_2^-$ | [218, 313] |
| II | $\overline{(CH_2)_3 COCLCO_2 Et} + F^-$ | [218, 313] |
| III | $2-MeC_6H_4COCH_2L + OH^-$ | [314] |
| IV | $4-MeC_6H_4COCH_2L + OH^-$ | [314] |
| V | $Ph_3C.CLMeCN + Bu^tO^-$ | [315] |
| VI | $CH_3CL_2NO_2 + 2-Bu^t$-pyridine | [316] |
| VII | $Me_2CLNO_2 + 2,6$-dimethylpyridine | [158, 316–318] |
| VIII | $Me_2CLNO_2 + 2,4,6$-trimethylpyridine | [317–319] |
| IX | $PhCL_2NO_2 + OH^-$ | [162] |
| X | $PhCL_2NO_2 + HPO_4^{2-}$ | [162] |
| XI | $PhCL_2NO_2 + $ imidazole | [162] |
| XII | $PhCL_2NO_2 + $ morpholine | [162] |
| XIII | $Me(CH_2)_2CLNO_2CO_2Me + 2,4,6$-trimethylpyridine | [320] |
| XIV | $4-NO_2C_6H_4CL_2CN + EtO^-$ | [321] |
| XV | $4-NO_2C_6H_4CL_2NO_2 + Et_3N$ | [323, 325, 326] |
| XVI | $4-NO_2C_6H_4CL_2NO_2 + Bu_3N$ | [322] |
| XVII | $4-NO_2C_6H_4CL_2NO_2 + $ quinuclidine | [323, 325, 326] |
| XVIII | $4-NO_2C_6H_4CL_2NO_2 + PhC(:NH)NEt_2$ | [325] |
| XIX | $4-NO_2C_6H_4CL_2NO_2 + (NMe_2)_2C:NH$ | [324, 196, 326] |
| XX | $PhMeCLCH_2Br + EtO^-$ (elimination of $Br^-$) | [327, 328] |
| XXI | $PhCL_2CH_2Br + Bu^tO^-$ (elimination of $Br^-$) | [330] |
| XXII | $4-MeOC_6H_4CL_2CH_2Br + Bu^tO^-$ (elimination of $Br^-$) | [330] |

| | | |
|---|---|---|
| XXIII | $3-NO_2C_6H_4CL_2CH_2Br + Bu^tO^-$ (elimination of $Br^-$) | [330] |
| XXIV | $4-MeOC_6H_4CL_2CH_2Br + OH^-$ (elimination of $Br^-$) | [191, 194] |
| XXV | $4-MeOC_6H_4CL_2CH_2SMe_2{}^+ + OH^-$ (elimination of $Me_2S$) | [191, 194] |
| XXVI | $9-Br-9,9'-$bifluorenyl $+ Bu^tO^-$ (elimination of $Br^-$) | [329] |

*Hydrogen atom transfers*

| | | |
|---|---|---|
| XXVII | $CH_3CL_2OH + H$ | [331] |
| XXVIII | $CL_3OH + H$ | [165] |
| XXIX | $CL_3CO_2^- + H$ | [165] |
| XXX | $Me_2CLOH + H$ | [332] |
| XXXI | $n-C_7H_{15}L + CH_3{}^•$ | [333] |
| XXXII | $PhSL + Ph_3C^•$ | [334] |
| XXXIII | $2,4,6-tri-Bu^tC_6H_2OL + CH_3{}^•$ | [335] |
| XXXIV | $PhCL_3 + Bu^tO^•$ | [336] |
| XXXV | $C_6L_5CL_3 + 2,2,6,6$-tetramethylpiperidyl | [337] |
| XXXVI | $C_6L_{12} + 2,2,6,6$-tetramethylpiperidyl | [337] |
| XXXVII | $C_6H_5OL +$ styrylperoxy radical | [338] |
| XXXVIII | $2,6-di-Bu^tC_6H_3OL +$ styrylperoxy radical | [338] |
| XXXIX | $cumyl-O_2L +$ tetralylperoxy radical | [338] |
| XL | $1,2-C_6H_4(OL)_2 +$ polyvinylacetate radical (PVAR) | [168] |
| XLI | $1,2,3-C_6H_3(OL)_3 +$ PVAR | [168] |
| XLII | $2,4,6-Me_3C_6H_2OL +$ PVAR | [168] |
| XLIII | $2,3,4,6-Me_4C_6HOL +$ PVAR | [168] |
| XLIV | $2,3,5,6-Me_4-1,4-(OL)_2C_6 +$ PVAR | [168] |
| XLV | $3,4,5-(OL)_3C_6H_2-CMe_2CH_2CMe_3 +$ PVAR | [168] |
| XLVI | $2,5-Bu_2^t-1,4-(OL)_2C_6H_2 +$ PVAR | [168] |
| XLVII | $(3-Me-4-OL-C_6H_3-)_2CHMe_2 +$ PVAR | [168] |
| XLVIII | $(3-Bu^t-4-OL-C_6H_3-)_2CHMe_2 +$ PVAR | [168] |
| XLIX | $(2,6-Me_2-4-OL-C_6H_2-)_2CHMe_2 +$ PVAR | [168] |
| L | $(2,6-Pr_2^i-4-OL-C_6H_2-)_2CHMe_2 +$ PVAR | [168] |
| LI | $1,1,1',1'-Me_4-6,6'-(OL)_2-3,3'$-spirobi-indane $+$ PVAR | [168] |
| LII | $1,1,1',1',5,5',7,7'-Me_8-6,6'-(OL)_2-3,3'$-spirobi-indane $+$ PVAR | [168] |
| LIII | $CD_2:CH.CH:CH.CH_3 \rightarrow CHD_2.CH:CH.CH:CH_2$ ($k^H$) $CH_2:CH.CH:CH.CD_3 \rightarrow CH_2D.CH:CH.CH:CD_2$ ($k^D$) | [339] |
| LIV | $\overline{CL_3.CL.CL:CL.CL}:\overline{CL} \rightarrow CL_2:\overline{C.CL_2.CL:CL.CL_2}$ | [340] |
| LV | $2(MeCL_2)_2NO^• \rightarrow MeCL:N^+(CL_2Me)O^- + (MeCL_2)_2NOL$ | [341] |
| LVI | $2\overline{CL_2(CH_2)_2CL_2N}O^• \rightarrow \overline{CL_2(CH_2)_2CL:N^+} - O^-$ $+ \overline{CL_2(CH_2)_2CL_2NOL}$ | [341] |

*Oxidation reactions*

| | | |
|---|---|---|
| LVII | $PhCL(CF_3)OH + MnO_4^-$ | [319, 342] |
| LVIII | $CF_3CL(OH)O^- + MnO_4^-$ | [343] |
| LIX | $3,5-(NO_2)_2C_6H_3CL(CF_3)OH + CrO_3$ | [344] |
| LX | $PhCL_2OH + MnO_2$ | [345] |
| LXI | $(4-Me_2NC_6H_4)_3CL +$ chloranil | [211] |

Table 5.3. *Anomalously large hydrogen isotope effects in solution reactions*

For key to reactions and references see Table 5.2

| Reaction | Solvent | $T(K)$ | $k^H/k^D$ |
|---|---|---|---|
| *Proton transfers* | | | |
| V | Bu$^t$OD | 298 | 15 |
| VII | H$_2$O | 298 | 19 |
| VII | Bu$^t$OH/H$_2$O | 298 | 24 |
| VIII | Bu$^t$OH/H$_2$O | 298 | 23[a] |
| XIII | H$_2$O | 303 | 27 |
| XV | PhCH$_3$ | 298 | 11 |
| XV | PhCl | 298 | 23 |
| XV | PhOCH$_3$ | 298 | 23 |
| XV | CH$_3$CN | 298 | 3.1[b] |
| XVI | PhCH$_3$ | 298 | 14 |
| XVI | CH$_3$CN | 298 | 2.2[b] |
| XVII | PhCH$_3$ | 298 | 16 |
| XVII | PhCl | 298 | 21 |
| XVII | PhOCH$_3$ | 298 | 19 |
| XVIII | PhOCH$_3$ | 298 | 32[b] |
| XIX | cyclohexene | 298 | 33[c,d] |
| XIX | mesitylene | 298 | 31[c,d] |
| XIX | PhCH$_3$ | 298 | 45[c,d] |
| XIX | Bu$_2$O | 298 | 41[c] |
| XIX | PhOCH$_3$ | 298 | 36[c] |
| XIX | PhCl | 298 | 50[c] |
| XIX | tetrahydrofuran | 298 | 13[c] |
| XIX | CH$_2$Cl$_2$ | 298 | 11[c] |
| XIX | CH$_3$CN | 298 | 12[c] |
| *Hydrogen atom transfers* | | | |
| XXVII | H$_2$O | [e] | 17 |
| XXVIII | H$_2$O | [e] | 20 |
| XXIX | H$_2$O | [e] | 22 |
| XXXIII | C$_7$H$_{14}$ | 363 | [f] |
| XXXV | PhCL$_3$ | 263 | 24 |
| XXXVI | C$_6$L$_{12}$ | 295 | 16 |
| XXXVII | styrene | 338 | 15 |
| XXXVIII | PhCl | 303 | 30 |
| XXXIX | PhCl | 303 | 17 |
| XL | MeCO$_2$CH:CH$_2$ | 323 | 13 |
| XLI | MeCO$_2$CH:CH$_2$ | 323 | 20[g] |
| XLII | MeCO$_2$CH:CH$_2$ | 323 | 14 |
| XLIII | MeCO$_2$CH:CH$_2$ | 323 | 17 |
| XLIV | MeCO$_2$CH:CH$_2$ | 323 | 14[g] |
| XLV | MeCO$_2$CH:CH$_2$ | 323 | 21 |
| XLVI | MeCO$_2$CH:CH$_2$ | 323 | 15 |
| XLVII | MeCO$_2$CH:CH$_2$ | 323 | 12[g] |

|       |                                    |     |      |
|-------|------------------------------------|-----|------|
| XLVIII| MeCO$_2$CH:CH$_2$                  | 323 | 13   |
| XLIX  | MeCO$_2$CH:CH$_2$                  | 323 | 15   |
| L     | MeCO$_2$CH:CH$_2$                  | 323 | 14   |
| LI    | MeCO$_2$CH:CH$_2$                  | 323 | 11   |
| LII   | MeCO$_2$CH:CH$_2$                  | 323 | 15   |
| LIII  | none                               | 298 | 12   |
| LV    | CF$_2$Cl$_2$                       | 298 | 14   |
| LVI   | C$_6$H$_6$                         | 298 | 13   |

*Oxidation reactions*

|       |                |     |      |
|-------|----------------|-----|------|
| LVII  | H$_2$O         | 298 | 16[h]|
| LVIII | H$_2$O         | 298 | 14   |
| LIX   | AcOH/H$_2$O    | 298 | 13   |
| LX    | heterogeneous  | [e] | 16   |

[a] Also $k^H/k^T = 79$.

[b] These values fall well within the classical limits, but have been included to illustrate the dependence of the isotope effect upon solvent.

[c] These results may need revision, since it has been found [346, 347] that under some conditions scrambling can take place between the CL$_2$-group in 4-nitrophenylnitromethane and the NH-group of amidines. It is unlikely, however, that the general picture will be greatly affected [348, 349].

[d] Marked positive deviations from the Arrhenius relation were observed at low temperatures for these reactions when R = H, also indicative of tunnelling.

[e] Temperature not stated, but probably near room temperature.

[f] $k^H/k^T = 50$.

[g] Temperature coefficients were measured for these reactions, but are not known sufficiently accurately to give reliable values for $E_A^D - E_A^H$ or $A_A^D/A_A^H$. However, the values of $A_A^D$ and $A_A^H$ are all the range $10^6$–$10^7$ dm$^3$ mol$^{-1}$ s$^{-1}$, which is $10^2$–$10^3$ times smaller than the values commonly found for hydrogen abstraction in the gas phase or in non-polar solvents: this is regarded by the authors [168] as further evidence for tunnelling.

[h] $k^H/k^T = 57$.

Table 5.4. *Anomalous hydrogen isotope effects on Arrhenius parameters for solution reactions*

For key to reactions and references see Table 5.2. Energies in kJ mol$^{-1}$

| Reaction | Solvent      | $E_A^D - E_A^H$ | $E_A^T - E_A^H$ | $A_A^D/A_A^H$ | $A_A^T/A_A^H$ |
|----------|--------------|-----------------|-----------------|----------------|----------------|
| *Proton transfers* | | | | | |
| I        | D$_2$O       | 6               | 14              | 2.9            | 25             |
| II       | D$_2$O       | 10              | 11              | 24             | 50             |
| III      | H$_2$O       | —               | 13              | —              | 30             |
| IV       | H$_2$O       | —               | 7               | —              | 13             |
| VI       | EtOH/H$_2$O  | 6               | —               | 11             | —              |
| VII      | EtOH/H$_2$O  | 6               | —               | 14[a]          | —              |

| | | | | | |
|---|---|---|---|---|---|
| VIII | BuOH/H$_2$O | 6 | — | 11 | — |
| IX | H$_2$O | 5 | — | 2.4 | — |
| X | H$_2$O | 6 | — | 5.2 | — |
| XI | H$_2$O | 5 | — | 2.1 | — |
| XII | H$_2$O | 6 | — | 10 | — |
| XIV | EtOH/Et$_2$O | 8 | — | 5 | — |
| XV | PhCH$_3$ | 12 | — | 4 | — |
| XV | PhCl | 16 | — | 25 | — |
| XV | CH$_3$CN | 4 | — | 1.8 | — |
| XVI | PhCH$_3$ | 7 | — | 1.1[b] | — |
| XVI | CH$_3$CN | 3 | — | 5 | — |
| XVII | PhCH$_3$ | 12 | — | 9 | — |
| XVII | PhCl | 11 | — | 4 | — |
| XIX | cyclohexene | 23[c] | — | 260[c] | — |
| XIX | mesitylene | 20[c] | — | 87[c] | — |
| XIX | PhCH$_3$ | 18[c] | — | 32[c] | — |
| XIX | Bu$_2$O | 18[c] | — | 27[c] | — |
| XIX | PhCl | 15[c] | — | 11[c] | — |
| XIX | tetrahydrofuran | 8[c] | — | 1.6[c] | — |
| XIX | CH$_2$Cl$_2$ | 8[c] | — | 2.2[c] | — |
| XIX | CH$_3$CN | 6[c] | — | 1.0[b,c] | — |
| XX | EtOH | 8 | 13 | 2.5 | 3.0 |
| XXI | Bu$^t$OH | 7 | — | 2.2 | — |
| XXII | Bu$^t$OH | 10 | — | 6 | — |
| XXIII | Bu$^t$OH | 8 | — | 2.2 | — |
| XXIV | Me$_2$SO/H$_2$O | 10 | — | 7 | — |
| XXV | Me$_2$SO/H$_2$O | 10 | — | 8 | — |
| XXVI | Bu$^t$OH | 11 | — | 10 | — |

*Hydrogen atom transfers*

| | | | | | |
|---|---|---|---|---|---|
| XXX | 6 M H$_2$SO$_4$ | 8[d] | — | 2.4 | — |
| XXXI | C$_7$H$_{16}$ | — | 14 | — | 5.0 |
| XXXII | PhCH$_3$ | — | 11 | — | 5.3 |
| XXXIV | PhCl | 11[e] | — | 30[e] | — |
| LIV | CCl$_4$ | 10 | — | 10 | — |

*Oxidation reactions*

| | | | | | |
|---|---|---|---|---|---|
| LVII | H$_2$O | 10 | — | 3.0 | — |
| LVIII | H$_2$O | 8 | — | 2.4 | — |
| LXI | CH$_3$CN | 14 | — | 24 | — |

[a] The authors [316] give $A_A^D/A_A^H = 7$: the value in the table is calculated from their reported values of $k^H/k^D$.

[b] These values are within the semi-classical limits, but are included to illustrate the effect of solvent.

[c] See Note c to Table 5.3.

[d] The authors [332] report a different value for the difference in activation energies, defined by $\Delta E = RT \ln (k^H/k^D)$, and hence based on the assumption that $A^H = A^D$.

# Experimental evidence for tunnelling

$k^H/k^D = 13$, 15 and 18 for different bases at 293 – 303 K, and these high values were accounted for by arguments similar to those given above.*

Large isotope effects can also arise in the overall rate of chain reactions for reasons unconnected with the tunnel effect. This has been observed in the autoxidation of organic substances by atmospheric oxygen: for example $k^H/k^D = 17$ at 373 K for $m$-xylene, 21 at 343 K for benzyl $t$-butyl ether, and 76 at 338 K for cumene [311, 312]. These figures refer to the maximum rate of oxygen consumption, and a simplified treatment of the generally accepted reaction scheme shows that for the autoxidation of a substance RL this maximum rate is given by

$$-\left(\frac{d[O_2]}{dt}\right)_{max} = \frac{k_1^2}{2k_2}[RL]^2 \qquad (5.4)$$

where $k_1$ and $k_2$ refer respectively to the reactions

$$RO_2^{\cdot} + RL \rightarrow RO_2L + R^{\cdot}$$

and

$$2RO_2^{\cdot} \rightarrow RO_2R + O_2.$$

Of these only the first will exhibit a considerable isotope effect, and since $k_1$ occurs to the second power in Equation 5.4 the occurrence of large isotope effects is understandable.

## 5.5 Comments on Tables 5.3 and 5.4

These tables provide ample evidence of deviations from semi-classical behaviour in the direction characteristic of tunnelling. Most of the reactions considered involve the breaking of carbon–hydrogen bonds, and in these instances the semi-classical upper limit for $k^H/k^D$ at 298 K has been taken as 10. Corresponding values for $k^H/k^T$, for other temperatures and for other bonds are given in Table 4.2 (p. 83). As emphasized previously, because of the presence of isotopically sensitive zero-point energy in the transition state, values of $k^H/k^D$ or $E_A^D - E_A^H$ which fall within the semi-classical limits may well be consistent with the presence of a considerable tunnel correction. This accounts for the fact that in Table 5.4 several of the reactions having markedly high values of $A_A^D/A_A^H$ nevertheless have values of $E_A^D - E_A^H$ which are little (if at all) greater than the semi-classical maximum: similarly, some of the reactions listed in Table 5.4 do not exhibit large enough isotope effects to be

---

* The expressions given by Ahlberg are actually more complicated than Equation 5.3, since the conversion of either isomer into the ion pair is assumed to be reversible, but the same principle is involved.

included in Table 5.3. A value of $A_A^D/A_A^H$ appreciably greater than unity is thus the best single criterion for tunnelling.*

The accuracy of the figures in Tables 5.3 and 5.4 is often difficult to estimate, and no attempt has been made to assign individual error limits. The values of $k^H/k^D$ or $k^H/k^T$ should usually be accurate to within a few per cent, while a reasonable assessment of the average error in the difference in activation energies is $\pm 1$ kJ mol$^{-1}$, corresponding to $\pm 0.17$ in $\log(A_A^D/A_A^H)$ at 300 K. The effects ascribed to tunnelling are mostly well outside these error limits, and it is particularly significant that all the values of $A_A^D/A_A^H$ and $A_A^T/A_A^H$ are greater than unity, as expected for moderate degrees of tunnelling. It is in fact difficult to find any well-documented instance of the opposite behaviour. It has been reported [350, 351] that $A_A^D/A_A^H$ has the values $0.29 \pm 0.12$, $0.53 \pm 0.37$ and $0.58 \pm 0.29$ for the reaction of di-(4-nitrophenyl)-methane with $t$-butoxide ions in various mixtures of $t$-butanol with toluene, while the reaction of phenylnitromethane with piperidine in aqueous solution [162] gave $A_A^D/A_A^H = 0.33 \pm 0.17$. Many other reactions of this type give values of $A_A^D/A_A^H$ considerably greater than unity (cf. Table 5.4), and there seems no reason why these two should be exceptional. A careful study of the oxidation of $HCO_2^-$ and $DCO_2^-$ by permanganate in aqueous solution [352] gave $A_A^D/A_A^H = 0.8$, i.e. close to the semi-classical value. Proton abstraction by acetate ions from phenyl $sec$-butyl ketone [353] also gave $A_A^D/A_A^H = 1.0$: in this reaction the great endothermicity would preclude any considerable tunnel effect. Similarly, $A_A^D/A_A^H = 0.9$ has been reported for the elimination reaction of 1,1-diphenylethane-2-benzenesulphonate in the presence of methoxide ions [354]. Apart from these few exceptions it appears that values of $A_A^D/A_A^H$ or $A_A^T/A_A^H$ greater than unity are the general rule for reactions in solution involving the transfer of protons or hydrogen atoms, suggesting that moderate tunnel corrections are frequently present.

The above considerations are all qualitative, and a more quantitative treatment would clearly be desirable. In view of the difficulties encountered for simple gas reactions and outlined in Sections 5.3.1 and 5.3.2 it is clearly impracticable to attempt a complete theoretical treatment for relatively complicated reactions in solution, especially since the experimental results are often not very extensive or accurate. It is, however, possible to show that the observed behaviour can be accounted for by tunnel corrections based on barriers of reasonable dimensions, and this approach has been adopted especially by Caldin and his collaborators [195, 196, 321–326, 354]. It is usual to assume that the barrier is parabolic and one-dimensional, that its form is

---

* As shown is Section 4.3.1 it is possible under some conditions for a large tunnel effect to be accompanied by values of $A_A^D/A_A^H$ equal to or less than unity. However, this will always entail a very large isotope effect and hence a low temperature. This situation applies to most of the reactions described in Section 5.2, but does not arise for the material presented in this section.

# Experimental evidence for tunnelling

Table 5.5. *Dimensions of parabolic barriers derived from hydrogen isotope effects*

For key to reactions and references see Table 5.2. $a$ = width of barrier; $E^H$, $E^D$ = barrier heights in kJ mol$^{-1}$

| Reaction | Solvent | $a$(pm) | $v_{\ddagger}^H$ (cm$^{-1}$) | $E^D - E^H$ | $E_A^H/E^H$ | $Q_t^H/Q_t^D$ |
|---|---|---|---|---|---|---|
| I | D$_2$O | 117 | 924 | 1.8 | 0.88 | 1.79 |
| II | D$_2$O | 117 | 1106 | 2.5 | 0.81 | 2.45 |
| VIII | BuOH/H$_2$O | 114 | 965 | 3.8 | 0.88 | 2.76 |
| XIV[a] | EtOH/Et$_2$O | 163 | 600 | 4.2 | 0.89 | 1.89 |
| XX[b] | EtOH | 159 | 908 | 3.8 | 0.95 | 1.74 |
| XIX | CH$_3$CN | 96 | 890 | 4.8 | 0.85 | 1.56 |
| XIX | CH$_2$Cl$_2$ | 95 | 936 | 4.8 | 0.84 | 1.65 |
| XIX | tetrahydrofuran | 90 | 998 | 4.8 | 0.89 | 1.75 |
| XIX | PhCl | 79 | 1388 | 5.0 | 0.49 | 5.8 |
| XIX | Bu$_2$O | 78 | 1408 | 4.8 | 0.50 | 6.0 |
| XIX[a] | PhCH$_3$ | 79 | 1420 | 4.8 | 0.49 | 6.5 |
| XIX[a] | mesitylene | 80 | 1540 | 2.1 | 0.43 | 13 |
| XIX | cyclohexene | 82 | 1563 | 1.0 | 0.41 | 22 |

[a] Low-temperature deviations from the Arrhenius equation lead to very similar barrier dimensions.

[b] Results for the reaction with tritium yield identical barrier dimensions.

independent of isotopic mass,* and that the reduced mass is equal to the mass of the appropriate hydrogen isotope. If it is further assumed that the semi-classical pre-exponential factors are independent of isotopic mass it is possible to use computer programs to calculate the barrier characteristics which best reproduce the observed isotope effects over a range of temperatures. The quantities which then emerge are the barrier heights for the two isotopes, $E^H$ and $E^D$ (which differ because of the mass-dependent zero-point energies in the initial and transition states), and either the barrier width or its curvature at the top. For a given type of barrier, such as the parabola, the last two quantities are not independent, and the curvature can conveniently be expressed as the imaginary frequency $iv_{\ddagger}$ for one of the isotopes.

Table 5.5 gives examples, which could readily be multiplied, of barrier dimensions calculated for some of the reactions in Tables 5.2, 5.3 and 5.4.

* In some calculations it has been assumed that the width of the barrier at its base is the same for two isotopes. This assumption is not strictly valid because of the differing heights of the zero-point levels, and it implies that the curvature of the barrier differs slightly for the two isotopes. It is more logical to start with the assumption of equal curvatures, leading to slightly different widths: fortunately the final results of the two procedures are essentially the same [324]. Similarly, many calculations have been based on a symmetrical truncated parabola, which assumes that the enthalpy of reaction (which is not always known) is zero. The results are of course affected quantitatively by whether the barrier is assumed to be symmetrical or unsymmetrical, but the general picture which emerges is the same.

Several points of interest emerge from the values in Table 5.5, as follows:

(a) The barrier widths $a$ all have reasonable values in the region of 100 pm, i.e. somewhat less than a bond length. Because of the artificial shape of the truncated parabola near its base the actual distance through which the proton moves will be rather greater than $a$.

(b) The values of the imaginary frequency $v_{\ddagger}$ are close to those predicted by various models of proton-transfer reactions [102, 105, 151, 177] and also to those derived from theoretical energy surfaces for hydrogen atom transfers (cf. Table 5.2).

(c) The difference between the barrier heights, $E^D - E^H$, is always less than the difference between the zero-point energies of the initial state (about 6 kJ mol$^{-1}$; cf. Table 4.2). This contrasts with the higher values frequently found for the difference between the Arrhenius activation energies, $E_A^D - E_A^H$, shown in Table 5.4.

(d) The last two columns of Table 5.5 indicate the extent to which the reactions deviate from classical behaviour. $E_A^H$ is a measure for the light isotope of the average excess energy of the systems which react, and the table shows that for the first eight entries (and for most of the other reactions in Table 5.2) it amounts to 80–95% of $E^H$, the barrier height. This implies that for most reactions at ordinary temperatures the classical picture of proton transfer is not radically altered by the inclusion of tunnel corrections. Correspondingly, for these systems the effect of tunnelling on the isotope effect, measured by $Q_t^H/Q_t^D$, is not drastic, though quite substantial. The last four entries in the table show much greater deviations from classical behaviour, in that for the light isotope the average energy of the reacting systems is now only 40–50% of the barrier height, and the tunnel correction $Q_t^H/Q_t^D$ now produces abnormally large isotope effects. The individual tunnel corrections are also quite large: for example, for the last entry $Q_t^H = 130$ and $Q_t^D = 6$.

(e) It is satisfactory that the investigations of H/D and H/T isotope effects leads to identical barrier dimensions for reaction XX.* As indicated in the notes to Table 5.5, for three of the reactions deviations from the Arrhenius equation at fairly low temperatures for the light isotope lead to essentially the same barrier dimensions as those based on the deuterium isotope effect. The reactions of the 2,4,6-trinitrobenzyl anion with acetic acid [356] and with hydrogen fluoride [357, 358] have been investigated in ethanol down to 159 K and 183 K, respectively. Although it was not possible to measure isotope effects both reactions showed positive deviations from the Arrhenius equation in the lower part of the temperature range, and analysis of these gave barrier dimensions similar to those in Table 5.5. These results have an obvious affinity with the studies of hydrogen transfers down to much lower temperatures described in Section 5.2.

* A similar comparison for Reactions I and II [218, 313] shows qualitative consistency, but there are some quantitative anomalies.

In spite of the qualitative success of calculations such as the above, not too much quantitative importance should be attached either to the absolute values of the barrier dimensions obtained, or to their variation from one system to another. However, one type of variation certainly merits further consideration, namely the dependence of the tunnel effect on the solvent, which appears most strikingly in the results of Caldin and Mateo [196] for Reaction XIX (4-nitrophenylnitromethane + tetramethylguanidine) in eight different solvents. Change of solvent produces large variations in the magnitude of the isotope effects, in the isotope effects on the Arrhenius parameters, and hence in the derived barrier dimensions, notably the barrier frequency $v_{\ddagger}^{H}$. There is no clear correlation of these quantities with any physical properties of the solvents, but it is noticeable that the latter fall into two groups. In the first group (acetonitrile, dichloromethane and tetrahydrofuran) the tunnel corrections are relatively small and $v_{\ddagger}^{H}$ is 900–1000 cm$^{-1}$, while in the second group (chlorobenzene, dibutyl ether, toluene, mesitylene and cyclohexene) the tunnel corrections are large and $v_{\ddagger}^{H}$ is 1400–1600 cm$^{-1}$: similarly, the barrier widths are about 95 pm in the first group and 80 pm in the second.

If the effective masses for tunnelling are independent of solvent, as has been assumed in the calculations so far, then the difference between the values of $v_{\ddagger}^{H}$ in the two groups of solvents implies a two-fold difference in the barrier curvatures, which seems physically unreasonable. It seems much more likely that a change of solvent can affect the effective tunnelling mass by coupling between proton motion and solvent motion, in particular the rotation of dipolar solvent molecules. The process of proton transfer involves a considerable charge displacement, and in the reaction concerned, in which an ion pair is formed from two uncharged molecules, the transition state is much more polar than the initial state: thus coupling with solvent dipoles seems very probable. This coupling can be represented by a simple molecular model [196, 209] and is also implicit in an electrostatic treatment of proton-transfer reactions [359]. The picture is that in the highly polar solvents of the first group coupling with solvent dipoles increases the effective mass of the proton sufficiently to reduce the manifestations of tunnelling markedly. In the second group of less polar solvents the proton–solvent interaction is mainly through electron polarization, which will not entail any appreciable increase in the reduced mass, so that the tunnel corrections remain large. The quantitative aspects of these ideas are discussed further in Section 7.2.4.

The same type of solvent effect has been observed in the reaction of 4-nitrophenylnitromethane with other nitrogen bases, but there is no information on the influence of the solvent on isotope effects in other proton-transfer reactions, although studies of this kind would be of great interest and may provide a new route for exploring the involvement of the solvent in such processes. If the picture presented above is a valid one the solvent should have little effect on tunnel corrections for hydrogen atom transfers, since these involve little change in charge distribution, and this is in fact the case in the few instances where the effect of solvent has been studied [231–233, 145, 448].

In spite of many speculations little progress has been made in discovering which structural features of the reactants are particularly favourable to tunnelling. The only generalization which emerges is a tendency for the presence of bulky groups in either or both reactants to give rise to substantial tunnel corrections, as may be seen from Tables 5.2–5.4. This was first pointed out by Lewis [317], who suggested that the role of steric hindrance was to hold the reactants apart, thus producing a high barrier with a large curvature at the top. However, this does not necessarily follow, since the result depends on the details of the potential energy curves involved. If the transition state is determined by the intersection of two parabolae and the only effect of steric hindrance is to move these further apart, then they will certainly intersect at a more acute angle and an enhanced tunnel correction may be expected. If on the other hand we adopt the electrostatic model described in Section 3.6.2, Equation 3.53 shows that an increase in $2L$, the distance apart of the two negative centres, will produce a decrease in the barrier frequency $v_{\ddagger}$, and hence a decrease in the tunnel correction. An alternative interpretation, at least for polar solvents, suggests that the effect of the bulky groups is to keep solvent molecules away from the seat of reaction and hence to reduce the increase in the effective mass of the proton produced by coupling with the solvent. It is interesting to note that in the racemization of $Ph_3C^\cdot CLMeCN$ by potassium $t$-butoxide in $t$-butanol (Reaction V) the isotope effect of $k^H/k^D = 15$ at 298 K is reduced to 5 by the addition of crown ether [315]. This can be explained by supposing that in the former case the potassium ion forms an ion pair with the transition state, thus effectively excluding solvent molecules from the already crowded system, while the addition of crown ether removes the potassium ion and permits the approach of a solvent molecule, thus increasing the effective mass of the proton. However, this (like much of the last two paragraphs) is largely speculation, and any firm interpretation of the effects of structure and solvent upon tunnel corrections must certainly await more experimental work and a better understanding of the detailed mechanism of proton-transfer reactions.

### 5.6 Tunnelling by entities other than hydrogen

Apart from electrons, the only particles usually regarded as candidates for tunnelling in chemical reactions are the isotopes of hydrogen. This is of course on account of their low mass, but if experiments can be carried out at sufficiently low temperatures tunnelling by much heavier entities may be expected. This can be seen by considering the parameter $u_{\ddagger} = hv_{\ddagger}/\mathbf{k}T$ which occurs in the expressions given in Chapter 3 for tunnel corrections for a parabolic barrier. For a given barrier $v_{\ddagger} \propto m^{-1/2}$, where $m$ is the mass of the

# Experimental evidence for tunnelling

particle, so that if a tunnel correction of a particular magnitude is observed for the light isotope of hydrogen at a temperature $T_H$, under similar circumstances the same tunnel correction should be observed for an entity of molecular mass $M$ at a temperature $T_H/M^{1/2}$. The same point can be expressed in terms of the characteristic temperature $T_c$ introduced by Christov [66, 99–101] and defined in Equation 3.43. $T_c$, which is also proportional to $M^{-1/2}$, is the temperature at which the reacting systems are approximately equally divided between those with $W > V_0$ and those with $W < V_0$, and at which the tunnel correction $Q_t$ is equal to $\pi/2$.

Many reactions involving the transfer of hydrogen atoms or protons exhibit appreciable tunnel corrections at 300 K and very striking effects below 100 K, and we should therefore expect similar behaviour by atoms or groups of mass 30 at temperatures of about 60 K and 20 K respectively. Since it is not possible to produce drastic changes in mass by isotopic substitution, evidence for tunnelling by heavy entities is to be sought in large deviations from the Arrhenius equation, including the possibility that the reaction velocity may become independent of temperature at sufficiently low temperatures. Studies of reaction velocities at very low temperatures have been recently reviewed by Goldanskii [360, 361], and there are at least two reactions for which such evidence appears to be well established.

The first is the growth of polymer chains in solid formaldehyde at temperatures between 4.2 K and 150 K, studied by Goldanskii and his collaborators [362–369]. Polymerization was initiated by $\gamma$-irradiation, and the growth of the polymer chains was followed calorimetrically either during irradiation or after it was cut off. The propagation reaction is

$$\text{-\!\!\!\sim\!\!\!\sim\!\!\!-} CH_2O^{\cdot} + CH_2O \rightarrow \text{-\!\!\!\sim\!\!\!\sim\!\!\!-} CH_2 - O - CH_2O^{\cdot}$$

and the results were expressed in terms of $\tau$, the average time needed to add one more unit on to the polymer chain. Between 150 K and 80 K $\tau$ increases according to an Arrhenius law with an activation energy of about 10 kJ mol$^{-1}$, reaching a value of about $10^{-5}$ s at 80 K. Below 80 K, however, there are very large deviations from the Arrhenius equation, and at the five temperatures studied between 10 K and 4.2 K $\tau$ remains independent of temperature at about $10^{-3}$ s. As the authors point out, an extrapolation of the Arrhenius relation to low temperatures predicts values of $\tau$ of $10^{30}$ years and $10^{100}$ years at 10 K and 4.2 K respectively, so that the observation of any polymerization at all at these temperatures is a very striking phenomenon.

It was found [369] that replacement of hydrogen by deuterium has little effect on the rate of polymerization, and the authors therefore propose that the rate-determining step is the motion of a $CH_2O$ molecule as a whole towards the end of the chain, and that this is essentially a tunnelling process at low temperatures. Since the molecular weight of formaldehyde is 30, the occurrence of marked tunnelling below 80 K is consistent with the estimates made at the beginning of this section. The authors have shown that the limiting rate between 4.2 K and 10 K is quantitatively consistent with tunnelling from the lowest energy level, assuming a not unreasonable value for the distance through which the $CH_2O$ molecule has to move.*

The other class of reaction in which tunnelling of a heavy entity appears to be firmly established is the binding of carbon monoxide to the iron atom of heme proteins such as hemoglobin and myoglobin, studied by a group of workers in the U.S.A. The carbon monoxide is dissociated by a laser flash and recombination is followed spectrophotometrically with equipment which can cover times from $2 \times 10^{-6}$ s to $10^3$ s in a single sweep [370]. The earlier work with myoglobin [371, 372] covered the temperature range 40–320 K, but this was subsequently extended downwards to 2 K and to other heme proteins, notably $\beta$-hemoglobin [373].

At low temperatures the kinetics of a single reaction follow an unusual course: thus at 30 K an appreciable fraction of the $\beta$-hemoglobin molecules have reacted with CO after $10^{-5}$ s, and yet a considerable proportion of unreacted molecules remain after 500 s, so that the reaction can be followed over many orders of magnitude in time. This behaviour is attributed to 'polychromatic chemical kinetics', i.e. to the presence in the system of a broad spectrum of activation barriers rather than a single one [374, 375]. Kinetics of this kind have been observed previously in radiation-induced reactions in solids [376–78], where they are probably due to a variety of depths of the traps in which the active centres occur. In the present instance the spectrum of activation barriers is attributed to a variety of conformations of the tertiary structure of the polypeptide chain. Above about 230 K these conformations are in thermal equilibrium with one another, but at lower temperatures they become frozen in, thus giving rise to a range of activation barriers. Because of the form of the kinetic curves it is not possible to define a single velocity constant, and $t_{75}$, the time at which 75% of the carbon monoxide remains unbound, was used to characterize the rate. Above about 80 K $t_{75}$ gives a linear Arrhenius plot, but at lower temperatures this plot flattens out, and between 10 K and 2 K the rate becomes independent of temperature.

Just as in the polymerization of formaldehyde, these results can only be explained by tunnelling of a heavy species, in this case carbon monoxide ($M = 28$). If reasonable assumptions are made about the form of the spectrum of barrier heights it is possible to describe the results quantitatively in terms of tunnelling through barriers of variable height and a width of about 50 pm. Similar results were obtained for the binding of carbon monoxide to

---

* The distance deduced ($\approx 50$ pm) is rather less than would be expected on the basis of a model. This may be due to the rather drastic assumptions which have to be made in evaluating the transition probabilities, and it will be seen later (Section 7.2.4) that the method used by Goldanskii and his co-workers does sometimes involve the assumption of unreasonably short distances.

protoheme between 5 K and 290 K [379], and it is stated [373] that the same type of low-temperature behaviour has been observed in the binding of carbon monoxide by seven other heme proteins and of nitric oxide ($M = 30$) by myoglobin. In some instances, for example with myoglobin and protoheme, the position appears to be complicated by the existence of several successive activation barriers, but there can be little doubt about the general features of these reactions or their fundamental interpretation.

Interesting possibilities arise at the other end of the mass spectrum in the chemical reactions of *muonium*, which may be regarded as an isotope of hydrogen of mass 0.113, and which should therefore exhibit very marked effects of zero-point energy and tunnelling. Muonium ($\mu^+ e^-$, or Mu) is formed when the positive muon captures an electron from matter, and has the same energy states and ionization potential as the hydrogen atom. It can therefore undergo the same kinds of chemical reaction, which may produce a diamagnetic species, for example

$$I_2 + Mu \rightarrow MuI + I,$$

or a radical, for example

$$H_2O_2 + Mu \rightarrow MuO^{\cdot} + H_2O.$$

The muon decays with an average life of 2.2 $\mu$s, emitting a positron, so that only very fast reactions can be studied; however, the polarization of the positrons can be used to monitor the state of combination of the muon and hence to follow its reactions. It was thus possible to compare the rates of several reactions of Mu and H in solution [380], with the results listed in Table 5.6.

The reactions of both H and Mu with iodine and bromine have rates close to the diffusion-controlled limit, which will be the same for both species. The remaining reactions of H are appreciably slower, and for these Mu is more reactive by a factor of $10^2$–$10^3$, giving rates close to the diffusion-controlled value of $\approx 10^{11}$ dm$^3$ mol$^{-1}$ s$^{-1}$. In these reactions there are no mass-sensitive

Table 5.6. *Rates of reaction of hydrogen atoms and muonium (Mu) in solution*

| Reagent | $k^H$ | $k^{Mu}$ | $k^{Mu}/k^H$ |
|---|---|---|---|
| | (dm$^3$ mol$^{-1}$ s$^{-1}$) | | |
| $I_2$ | $4 \times 10^{10}$ | $6 \times 10^{10}$ | $\approx 1$ |
| $Br_2$ | $12 \times 10^{10}$ | $9 \times 10^{10}$ | $\approx 1$ |
| $H_2O_2$ | $9 \times 10^7$ | $1 \times 10^{10}$ | $10^2$ |
| $NO_3^-$ | $9 \times 10^6$ | $13 \times 10^{10}$ | $1.4 \times 10^3$ |
| $ClO_4^-$ | undetected | $4 \times 10^{10}$ | large |
| $Fe^{3+}$ | $9 \times 10^7$ | $2 \times 10^{10}$ | $2 \times 10^2$ |

vibrations in the reactants, so that any zero-point energy effects will lead to an inverse isotope effect ($k^H > k^{Mu}$) because of the presence of such vibrations in the transition state. This effect will be partly cancelled by the factor $(\mu_{\ddagger}^H/\mu_{\ddagger}^{Mu})^{1/2}$ in the expression for $k^{Mu}/k^H$ (cf. Equations 4.11 and 4.14), but the large observed values of $k^{Mu}/k^H$ can only be due to preferential tunnelling by muonium.

It has also been reported [381] that the reaction of muonium with bromine in the gas phase has a velocity constant of $(2.4 \pm 0.3) \times 10^{11}$ dm$^3$ mol$^{-1}$ s$^{-1}$ at 296 K. This is about ten times the average value reported for H + Br$_2$ and forty times that for D + Br$_2$, and the authors attribute these differences to tunnelling. However, the velocity constants for the last two reactions are not known with any certainty, and the values for all three reactions are fairly close to the gas collision numbers: the latter will be three times as great for Mu + Br$_2$ as for H + Br$_2$ because of the mass difference. These reactions have at most very low activation energies, so that large effects of tunnelling are not to be expected at ordinary temperatures.

# 6 Tunnelling in molecular spectra

## 6.1 Introduction

There are many features of molecular spectra which can only be explained by the same concepts as those on which the treatment of tunnelling problems is based, such as the uncertainty principle. Although it would be usually misleading to say that these features are 'caused' by tunnelling processes, they are sufficiently closely related to chemical reactions to merit a chapter in this book. There exists a vast literature on both the theoretical and the chemical aspects of this subject, and in many instances the complete picture is embellished with so much detail that it may well look rather obscure to the non-spectroscopic layman. No attempt will therefore be made either to cover the field completely or to deal adequately with spectroscopic niceties. Instead a simplified account will be given of a few selected topics which bear a fairly close analogy to chemical processes.

This analogy is probably closest for the predissociation spectra of simple species, particularly diatomic hydrides and deuterides, since the transitions concerned may result in the dissociation of the molecule, and this subject is discussed in Section 6.2. As for chemical reactions the most convincing evidence comes from a comparison of isotopic species. Section 6.3 is devoted to the spectroscopic consequences of the inversion of ammonia, since although this is not in itself of prime chemical interest it serves as a model for other systems in which the potential energy curve has a double minimum, such as hydrogen bonds and flexible ring systems: some of these are also considered in Section 6.3. Hindered rotation, for example of methyl groups, also involves a potential energy curve with a series of successive minima and therefore raises similar theoretical problems, but since the role of the tunnel effect in such systems has been mainly investigated by means of magnetic resonance spectra this subject is dealt with separately in Section 6.4. Finally, there are many systems in which the lifetime of a species or a configuration is reduced by the occurrence of kinetic processes other than those already mentioned, with consequent effects on the form of the magnetic resonance spectra. Observations of n.m.r. or e.s.r. spectra can therefore be used to measure the

## 6.2 The predissociation of diatomic hydrides and deuterides

The term predissociation is used to describe certain features of electronic spectra which are observed when the lifetimes of either the upper or the lower states concerned are reduced because of the possibility of molecular dissociation. Because of the uncertainty principle a short lifetime $\tau$ is associated with an uncertainty in energy given by $\Delta E \approx \hbar/\tau$, so that the effect is to broaden the spectral lines corresponding to transitions to or from the state concerned. If this broadening exceeds that due to other causes, such as pressure broadening or instrumental resolving power, it can be observed experimentally. If the broadening becomes too great the line may escape detection, since the available radiant energy is spread over too large a spectral range: moreover, the process which reduces the lifetime of a state will also deplete its population, leading to a reduction in the observed intensity due to transitions from this state. The signs of predissociation are therefore broadening, weakening or breaking off of a series of lines (usually rotational), and a lack of correspondence between the absorption and emission spectra of the same species, though not all these signs need be observable simultaneously.

The different ways in which predissociation spectra can arise have been clearly set out by Herzberg [382, 383] and elaborated by Mulliken [384]. In the present context the simplest case to consider is *rotational or centrifugal predissociation*. Consider a diatomic species of reduced mass $\mu$ and let the potential energy curve in the absence of rotation be $V_0(r)$: this will be a Morse-like curve such as the first curve in Fig. 6.1 with a minimum at a certain internuclear distance $r_e$, but no maximum: the first few vibrational levels have also been indicated on the diagram. If the molecule is also rotating, according to classical mechanics the equilibrium distance will be increased to a new value $r_c$ such that the centrifugal force is balanced by the restoring force, i.e.

$$\mu r_c \left(\frac{d\phi}{dt}\right)^2 - \frac{P^2}{\mu r_c^3} = V_0'(r_c) \tag{6.1}$$

where $\phi$ is the angle of rotation, $P$ the angular momentum and $V_0' = dV_0/dr$. For vibrations about the new equilibrium position the restoring force $F$ is given by

$$F = V_0'(r) - \frac{P^2}{\mu r^3} = \frac{d}{dr}\left[V_0(r) + \frac{P^2}{2\mu r^2}\right] \tag{6.2}$$

# Tunnelling in molecular spectra

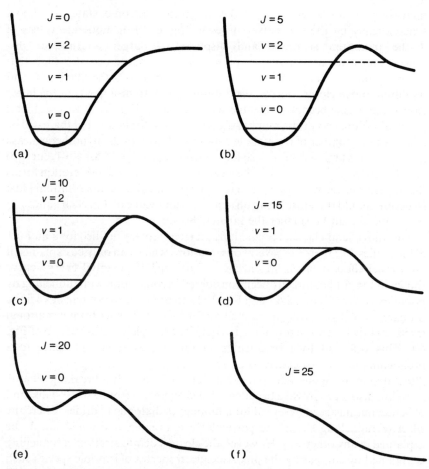

Fig. 6.1 Rotational predissociation. Schematic representation of the effect of rotational quantum number on the effective potential energy curve for dissociation

so that the *effective potential energy* $V(r)$ for the rotating molecule is

$$V(r) = V_0(r) + \frac{P^2}{2\mu r^2}. \tag{6.3}$$

Further, quantum theory shows that the angular momentum $P$ is equal to $\hbar[J(J+1)]^{1/2}$, where $J$ is the rotational quantum number. The effective potential energy for the $J$th rotational state is thus

$$V(r) = V_0(r) + \frac{\hbar^2}{2\mu r^2} J(J+1). \tag{6.4}$$

Equation 6.4 shows some interesting properties if we consider the effect of increasing $J$ for a given $V_0(r)$, i.e. a given electronic state, as shown schematically by the successive curves in Fig. 6.1. For moderate values of $J$ the minimum is raised and displaced to larger internuclear distances, and a maximum is now present with an energy greater than that corresponding to complete dissociation. As $J$ increases the maximum and minimum move closer together, and ultimately both disappear: in the latter case the molecule becomes unstable, as in curve f in Fig. 6.1. If we consider a given vibrational quantum number $v$ it is clear from the figure that there will be a limit to the number of observable rotational levels. In the hypothetical case illustrated this limit will be $J = 20$ for $v = 0$ (curve e), $J = 15$ for $v = 1$ (curve d) and $J = 10$ for $v = 2$ (curve c). If the electronic state under consideration forms the upper or lower state giving rise to a given band system, the rotational fine structure should therefore break off at a particular value of $J$ in each band, and this value should be smaller the greater the value of $v$.*

The above semi-classical predictions are qualitatively fulfilled for a number of molecules, but there are several observations which can only be explained if the tunnel effect is taken into account. The type of potential energy curve shown in Fig. 6.1 has already been considered in connection with tunnelling by bound particles (Section 2.6.1 and Fig. 2.11). In principle the system has a finite probability of dissociating by tunnelling through the barrier from vibrational levels slightly below the top of the barrier, for example $v = 2$ in curve b of Fig. 6.1. This will lead to a broadening of this energy level and hence to a broadening of the associated rotational lines. Thus in the hypothetical case illustrated in the figure we might expect for $v = 2$ not only a breaking off at $J = 10$ but also a progressive broadening as $J$ increases from 5 to 10. This kind of behaviour has been observed for a number of diatomic hydrides, where the effect of tunnelling should be particularly noticeable, and could hardly be explained in any other way. However, the clearest demonstration of tunnelling is obtained by comparing the predissociation spectra of isotopic species such as XH and XD. If we disregard tunnelling the energy of the last stable state will be equal to the maximum value of $V(r)$, and hence from Equation 6.4

$$E_{max} = V_0(r_{max}) + \frac{\hbar^2}{2\mu r_{max}^2} J_{max}(J_{max} + 1) \qquad (6.5)$$

in which the isotopic mass appears only in the factor $\mu$, since $V_0(r)$ is the same for the two isotopic species. If therefore the observed values of $E_{max}^H$ for different vibrational quantum numbers are plotted against $J_{max}(J_{max} + 1)$ to give the so-called limiting curve for predissociation, then the values of $E_{max}^D$

---

* The rotational selection rule for electronic transitions is $\Delta J = 0, \pm 1$, giving rise to P, Q and R branches in the spectrum, each of which should show the behaviour described above. When there is weak spin–orbit coupling the quantum number $J$ is replaced by $K$, which refers to the total angular momentum apart from spin, but the form of the above equations remains unchanged.

# Tunnelling in molecular spectra

should lie on the same curve provided that $J_{max}(J_{max}+1)$ is replaced by $J_{max}(J_{max}+1)\mu^H/\mu^D$.

On the other hand, if the tunnel effect is allowed for, a much greater effect of isotopic mass is predicted. A given degree of broadening (for example sufficient broadening to render the line in question undetectable) will be observed when the life-time falls to a certain value. As shown in Section 2.6.1 the life-time is given approximately by

$$\tau = \frac{1}{v}\exp\left\{\frac{2}{\hbar}\int[2\mu(V-E_{max})]^{1/2}\,dr\right\} \tag{6.6}$$

where $v$ is the classical vibration frequency and the integration is carried out between the two values of $r$ at which $V = E_{max}$. When comparing two isotopic species XH and XD the mass will affect $v$, giving a shorter life-time for the lighter isotope, but far more important is the factor $\mu^{1/2}$ in the exponent, which acts in the same direction. The result is that when the values of $E^H_{max}$ and $E^D_{max}$ are plotted against $J_{max}(J_{max}+1)$ and $J_{max}(J_{max}+1)\mu^H/\mu^D$, respectively, the points for the two isotopic species should not lie on a single curve, but the curve for XH should throughout be lower, corresponding to lower energies for predissociation because of the greater tunnelling ability of the lighter isotope. This has been verified for a number of diatomic hydrides. The first clear demonstration was by Olsson [385] for AlH and AlD, and recent examples include LiH and LiD [386] and HgH, HgD and HgT [387]. The latter papers also consider the reverse process to rotational predissociation, i.e. the combination of two atoms by a collision which is not 'head-on'. The effective potential energy curves for this last process are just the same as those illustrated in Fig. 6.1: for certain values of the angular momentum the system must pass over or through an energy barrier, and tunnelling is again important for light atoms.

A second cause of predissociation arises when the vibrational or rotational levels of an excited electronic state are overlapped by the dissociation continuum of another electronic state. This is illustrated in Fig. 6.2, where transitions between states I and II are affected by the possibility of radiationless transitions from state II to the repulsive state III. This process also involves the crossing of an energy barrier, with possibilities of tunnelling. However, the situation is less simple than for rotational predissociation, since apart from the possibility of tunnelling from energy states below the intersection, the broadening or absence of lines in the spectrum depends on the probability of transitions from state II to state III in the neighbourhood of the intersection.* Further, when rotational energy is taken into account the effective potential energy curve for state II may have a maximum, just as for purely rotational predissociation. Because of these complications we shall not consider this type of predissociation further.

---

* Because of the interaction between the states it is not strictly correct to picture a crossing of the two potential energy curves. Instead there is a splitting into an upper and a lower state, as indicated by the dotted lines in Fig. 6.2.

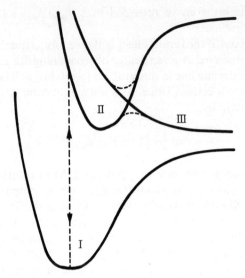

Fig. 6.2 Predissociation by radiationless transition

Although the electronic spectra of polyatomic molecules or radicals are often difficult to analyse completely, there are several instances in which hydrides exhibit a fairly diffuse spectrum which is considerably sharpened if hydrogen is replaced by deuterium, and this effect must be attributed to a greater degree of tunnelling by the lighter isotope. Such behaviour has been observed for water [388], ammonia [389] and the methyl and methylene radicals [390]. It is interesting to note that in the case of methylene the spectrum of $CH_2$ was too diffuse to permit conclusions about the geometry of the radical which had to be deduced from the spectrum of $CD_2$.

In small molecules radiationless transitions between electronic states (for example between states II and III in Fig. 6.2) can take place at or near the intersections of potential energy curves or surfaces. In large molecules these surfaces exist in space of many dimensions, so that such intersections are few, and it has been suggested by a number of authors [392–394] that radiationless transitions in such systems may involve the tunnelling of hydrogen atoms between the surfaces concerned. A highly simplified quantitative treatment of the process has been given [395] and some experimental support is claimed, but at present the suggestion must be regarded as speculative.

## 6.3 The inversion of ammonia and related processes

The inversion of ammonia has already been mentioned in Section 2.6.2 as the prototype of processes involving tunnelling in a symmetrical potential energy curve with two minima separated by a maximum. Instead of the broadening observed in predissociation this situation causes the vibrational levels to be split, the separation of the doublet being given approximately by Equation

# Tunnelling in molecular spectra

2.115. This splitting appears experimentally in the vibration–rotation spectrum of ammonia, first explained theoretically by Dennison and Uhlenbeck [71]. The topic has been reviewed in detail in several books on spectroscopy [396–398], and for this reason only a few of the main results will be mentioned here, without detailed literature references.

The simplest situation arises in the pure rotational spectrum, observable in the infra-red. The selection rules for the infra-red demand that $\Delta J = \pm 1$, where $J$ is the quantum number for rotation about an axis perpendicular to the symmetry axis, and also that transitions take place between the upper component of one doublet and the lower component of the other. Since the splitting is essentially the same for two adjacent rotational levels the observed splitting in the pure rotational spectrum is thus just twice the doublet separation in the lowest vibrational state. The observed splitting for $NH_3$ is 1.32 cm$^{-1}$, so that the doublet separation is about 0.7 cm$^{-1}$. Turning now to the vibrational spectrum, an infra-red band must again correspond to transitions between the upper component of one doublet and the lower component of the other, so that the observed separation will be equal to the sum of the splittings in the two vibrational levels. Since we already know the splitting of the lowest vibrational level, this makes it possible in principle to deduce the splitting of higher levels.*

In the double-minimum energy diagram as usually drawn (Fig. 2.12) the co-ordinate $x$ represents the distance of the nitrogen atom from the plane of the three hydrogens, i.e. the height of the pyramid. This type of motion does not correspond exactly to any of the normal modes of vibration, but one would expect the effect of tunnelling to be most marked in those vibrations which are accompanied by a considerable change in the height of the pyramid. This condition is satisfied by the two non-degenerate vibrations $v_1 = 3337$ cm$^{-1}$ and $v_2 = 950$ cm$^{-1}$ (approximately symmetrical stretching and symmetrical bending respectively) illustrated in Fig. 6.3. Since the ammonia molecule is rather a flat pyramid the greatest effect should be observed for $v_2$. This is in fact what is observed, the doublet splitting for the first excited state ($v_2 = 1$, other vibrational quantum numbers zero) being 36 cm$^{-1}$, while the corresponding splitting for $v_1 = 1$ is only 0.8 cm$^{-1}$. The increase of splitting with vibrational quantum number is of course in accordance with Equation 2.115.

More direct and much more accurate information about inversion doubling comes from microwave spectroscopy. The selection rules show that transitions between the two levels of a doublet are active for the absorption or emission of radiation, and the splitting of the lowest level of $NH_3$ can therefore be observed directly in the centimetre wavelength region. This observation was first made in 1934 by Cleeton and Williams [399], and a monograph published

---

* The position is somewhat different for the Raman spectrum, since the selection rules are not the same.

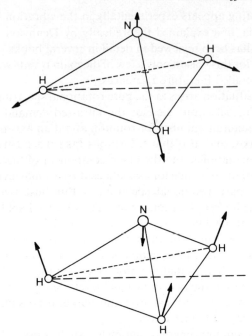

Fig. 6.3 Symmetrical vibrations of the ammonia molecule

30 years later [398] cites 105 references to the microwave spectrum of ammonia in the period 1945–63. The results of infra-red and Raman spectroscopy are confirmed, and the high resolving power of the microwave technique makes it possible to add much more detail. For example, it is found that the doublet separation is increased by rotation about the axis of the molecule, which flattens the pyramid and hence lowers the barrier, while the reverse effect is produced by rotation about a perpendicular axis, which sharpens the pyramid.

The frequency of the microwave radiation absorbed is of course given by $v = \Delta E/h$, where $\Delta E$ is the energy difference between the components of the doublet. We have already seen (Equation 2.119 and footnote) that the same expression is obtained for the tunnelling frequency, i.e. the frequency with which the system tunnels back and forth between the two potential wells, supposing that it has been initially located in one of them. However, not too much should be read into the equality of these two frequencies, and in particular it is not correct to regard the microwave absorption as being due to the passage of the system from one potential well to the other. Analogous statements apply to the vibrational spectrum of a harmonic oscillator, where (because of the selection rules) the frequency of the radiation absorbed or emitted is equal to the mechanical frequency of the oscillator, but is not directly related to it in a physical sense.

Inversion doubling in $ND_3$ has also been investigated thoroughly by a variety of spectroscopic techniques. In common with other manifestations of tunnelling there is a large effect of isotopic mass, and the splittings are throughout much smaller for $ND_3$ than for $NH_3$. In the lowest state the values are 0.05 and 0.8 cm$^{-1}$ respectively, while for the state with $v_2 = 1$ the corresponding values are 3.4 and 36 cm$^{-1}$. Information is also available for the species $NH_2D$ and $NHD_2$, which show intermediate values of the splitting, though their interpretation is complicated by the reduction in molecular symmetry. The experimental results available for the various isotopic forms of ammonia are more than sufficient to determine the parameters of the double-minimum potential, and the fact that a consistent picture can be obtained is good evidence for the validity of the theoretical treatment. The simplest approach is to assume some reasonable analytical form for the potential energy, for example that first proposed by Manning [72],

$$V(x) = A\,\text{sech}^4(x/2\rho) - B\,\text{sech}^2(x/2\rho) \tag{6.7}$$

and then to adjust the constants $A$, $B$ and $\rho$ so as to fit the observed frequencies and splittings. The BWK approximation (Equation 2.115) is usually sufficient for this purpose. The result, which depends only slightly on the analytical form assumed for the potential function, is that the maximum lies about 2000 cm$^{-1}$ (24 kJ mol$^{-1}$) above the two minima, which are separated by a distance of 76 pm. The last value agrees exactly with the molecular dimensions deduced from the rotational spectrum. On a classical basis, which assumes passage over the barrier, the frequency of inversion of $NH_3$ by the symmetrical bending vibration would be approximately $v_2 \exp(-24\,000/RT)$ i.e. about $10^9$ s$^{-1}$ at 300 K. The tunnelling frequency from the lowest vibrational level is 2.4 $\times$ 10$^{10}$ s$^{-1}$, so that tunnelling is predominant under these conditions. For $ND_3$ the tunnelling frequency falls to $1.6 \times 10^9$ s$^{-1}$ while the classical rate will be reduced by a factor of approximately $2^{1/2}$; thus even for this species tunnelling makes an important contribution at ordinary temperatures.

The same principles apply to the inversion of other pyramidal $XY_3$ molecules. For $PH_3$ the tunnel frequency from the lowest level is only $1.4 \times 10^6$ s$^{-1}$: although the height of the barrier is similar to that for $NH_3$ it is much wider, the distance between the two minima being 134 pm. Although tunnelling leads to a splitting of spectroscopic levels the actual process of inversion of $PH_3$ must be essentially classical. For $AsH_3$ it has been estimated that inversion by tunnelling takes place only about once every two years: this corresponds to a doublet splitting of less than $10^{-12}$ cm$^{-1}$, which is of course quite undetectable spectroscopically.

When a hydrogen bond is formed between two identical atoms there are nearly always two interconvertible configurations, $-X-H\cdots X- \rightleftharpoons -X\cdots H-X-$. The potential energy curve thus possesses two symmetrical minima, and the possibility of tunnelling should have spectroscopic consequences similar to those for ammonia. However, the experimental evidence appears to be rather confused, partly because many of the observations refer to the solid or liquid state, and partly because apparent splitting of infra-red frequencies may actually be due to overtones of bending vibrations [400]. Much of the evidence relates to the dimers of carboxylic acids, which have the following structures:

$$\begin{array}{c} \text{O—H---O} \\ R—C \diagup \qquad \diagdown C—R \\ \diagdown \text{O---H—O} \diagup \end{array}$$

Several authors [401–403] have invoked tunnelling to explain some features of the spectra of dimeric formic and acetic acids and their deuteriated forms, though others [404, 405] have doubted this interpretation. The transition between the two components of the split level should be observable in the far infra-red spectrum, and it has been claimed [406] that this has been achieved for hydrogen-bonded solids such as potassium dihydrogen phosphate, potassium hydrogen di(4-nitrobenzoate) and phenylphosphinic acid. Another pertinent investigation [407] deals with the infra-red spectrum of a dilute solid solution of HOD in ordinary ice. The O–H band is very broad – half-width about 300 cm$^{-1}$ – and much broader than the O–D band. This is attributed to tunnelling, and the fact that broadening is observed rather than splitting is explained by supposing that the O–H----O distance does not have a single fixed value but varies within a certain range. Similarly, the broad or continuous infra-red spectra of solutions or solid hydrates of strong acids has been explained in terms of tunnelling in hydrogen bonds [408, 409], though this interpretation has not been universally accepted. In general it seems extremely probable that tunnelling plays a part in the properties of hydrogen bonds and the problem has been treated theoretically by a number of authors [75, 76, 78, 79, 410–413], though definitive experimental evidence is rare.

## 6.4 Hindered rotations

When a methyl group in a molecule undergoes internal rotation through 360° the potential energy of the system passes through a number of successive maxima and minima, and the same is true when a species such as $NH_3$, $CH_4$ or $NH_4^+$ rotates in a crystal. If there is a finite probability of tunnelling between the orientations corresponding to successive minima the energy levels for torsional oscillation will be split in a manner which resembles that found for ammonia, but is usually much more complex. This splitting can be observed in infra-red and microwave spectra and has been frequently reviewed [414]. We shall not discuss this topic here, but instead shall consider the effect of rotation on magnetic resonance spectra at low temperatures, which in many instances shows that rotation is essentially a tunnelling process when the moment of inertia is low, as in $XH_3$ and $XH_4$ species. This subject has been recently reviewed by Srinivasan [415].

The spectra of most interest in the present context are proton magnetic resonance (p.m.r.) spectra. In solids or viscous liquids such spectra normally give broad lines because of the varying local fields due to neighbouring magnetic dipoles. In gases and in most liquids, on the other hand, these dipolar interactions are averaged to zero by the rapid tumbling of the molecule so that the lines become narrow and may also show fine structure owing to spin–spin interaction with other magnetic nuclei in the same molecule. Even in solids this narrowing may take place if either the molecule as a whole or the group containing the protons (commonly a $CH_3$ group) can rotate sufficiently freely at the temperature of the experiment: for example, the p.m.r. spectrum of solid methane gives narrow lines over a wide range of temperature [416, 417]. In this context 'sufficiently freely' means that the average time for rotation from one configuration to another shall be comparable with or less than the characteristic relaxation time of the p.m.r. spectrum which is commonly in the range $10^{-3}$ s to 1 s. On a classical basis rotation can only take place if the energy is sufficient to overcome the rotational barrier, the height of which can be estimated by various means, in particular from microwave spectra. For a given barrier the line shape can be calculated as a function of temperature, for example for rotating methyl groups [418]. The results show a continuous change from a narrow line (free rotation) at high temperatures to a broad line (no rotation) at low temperatures, and similar calculations can be made for tetrahedral species such as $CH_4$ and $NH_4^+$.

This classical picture is consistent with the experimental observations for some systems, but there are many instances in which a narrow-line spectrum persists down to very low temperatures at which the thermal energy is quite insufficient to overcome the barrier to rotation. For example, the signal for solid ammonia is narrower than expected in the range 2–50 K [419] and the same is true for solid methane down to 1.29 K [416]. This can only be due to rotation by tunnelling, and is confirmed by a quantitative theoretical treatment of the problem [420]. The theory also predicts that significant rotational tunnelling should cause the appearance of side-bands in p.m.r. spectra. These have been observed for solid $(CH_3)_2O$, $C_3H_8$ and $(CH_3)_2S$ [421] and are consistent with the theory and with the barrier heights calculated for the rotation of methyl groups in these molecules [422]. The effect of replacing $CH_3$ by $CD_3$ has also been considered [423], and interesting results were obtained for a dilute solid solution of $CH_3CD_2I$ in $CD_3CD_2I$ [424]. Between 40 K and 80 K the spectra correspond to tunnelling rotation of the methyl group, and at 87 K the spectrum indicates a fixed triangle of protons. Between 87 K and 127 K further changes take place on account of classical thermally-activated rotation.

The p.m.r. spectrum of many ammonium salts has been studied [425, 426] and is found to depend on the nature of the anion, since this determines the height of the barrier to rotation of the ammonium ion. With increase of temperature the spectrum of ammonium chloride shows a gradual decrease of line width which can be explained by the onset of classical rotations. Most ammonium salts, on the other hand, give evidence of rotational tunnelling. This has been interpreted theoretically [427–429], and in particular a relation has been found between the tunnelling frequency and the barrier height.

Similar considerations apply to the effect of the rotational tunnelling of groups or molecules in electron spin resonance spectra [430–432], though the time scale involved is now shifted to the range $10^{-9}$–$10^{-5}$ s. Observations are of course restricted to species containing unpaired electron spins, and the subject will not be pursued here.

## 6.5 Magnetic resonance spectra in labile systems

The position in a magnetic resonance spectrum of the line due to a particular nucleus is affected by its environment (the so-called chemical shift). If a system contains magnetic nuclei of a particular kind, for example protons, in two different environments, the observed spectrum will depend on whether, and how rapidly, the nuclei can change places between the two sites, either by direct exchange or by exchange from each site with other species. If exchange is absent or very slow two separate sharp lines will appear for the two sites, while if exchange is very fast a single sharp line will appear at a position corresponding to a weighted mean of the two resonances. Intermediate rates of exchange will produce broader spectra, which may or may not have two peaks, and there will also be changes in fine structure. Observations in this intermediate range can therefore be used to determine the rate of the processes leading to exchange, and for proton magnetic resonance the accessible rates lie in the range $1$–$10^5 s^{-1}$. The theory and applications of this method have been described in many books and reviews [433–441]. We shall therefore regard this technique as an established method for measuring reaction velocities (though distinct from the use of magnetic resonance spectra as an analytical tool, as described in Section 5.1.), and shall mention only a few processes for which the results provide evidence for tunnelling.

Inversion about nitrogen in aziridine (sometimes called ethyleneimine) is formally similar to the inversion of ammonia, though it is very much slower because two of the valencies of nitrogen are constrained in the ring. The process involved is shown schematically below.

# Tunnelling in molecular spectra

If the molecule is fixed in one of these configurations the four $CH_2$-protons fall into two pairs according to their distance from the NH-proton and each pair will have its own chemical shift in the p.m.r. spectrum, but if there is rapid inversion only a single line will be observed. Gas-phase p.m.r. measurements over the range 310–380 K have been used to determine the rate of inversion of aziridine and of aziridine-N-d [442]. The isotope effect is small ($k^H/k^D = 2.34$ at 338 K) because inversion is associated with an N–H bending frequency of only 1096 cm$^{-1}$ for the light isotope. However, the isotope effect on the Arrhenius parameters is found to be $E_A^D - E_A^H = 9$ kJ mol$^{-1}$, $A_A^D/A_A^H = 10$. Although these values are not very accurate they suggest a considerable tunnel correction: since only a bending mode is involved the semi-classical prediction would be $E_A^D - E_A^H < 1.7$ kJ mol$^{-1}$. Moreover, since all the fundamental vibration frequencies of aziridine are known [443] it is possible to make a fairly firm estimate of the semi-classical isotope effect on the basis of transition state theory. This yields values of $k^H/k^D$ which are much smaller than those observed: on the other hand the application of a tunnel correction calculated from Equation 3.23 with $v_t^H = 1050$ cm$^{-1}$ gives complete agreement with experiment. The tunnel correction $Q_t^H/Q_t^D$ varies from 2.16 at 308 K to 1.51 at 373 K and thus accounts for a large proportion of the observed isotope effect. The calculated barrier height is 80 kJ mol$^{-1}$, which is consistent with various theoretical estimates and compares with 24 kJ mol$^{-1}$ for the ammonia inversion. The observed activation energy for the light isotope is 63 kJ mol$^{-1}$, so that on all counts this process is characterized by moderate tunnelling, like most of the reactions listed in Tables 5.4 and 5.5.

A more striking example of tunnelling is provided by the inter-conversion in solution of the degenerate tautomers of meso-tetraphenylporphine, illustrated in Fig. 6.4. The term 'degenerate' signifies that the two forms are chemically identical, though the rate of their interconversion can be measured by n.m.r. spectroscopy. At low temperatures the aromatic protons marked $H_1$ and $H_2$ in

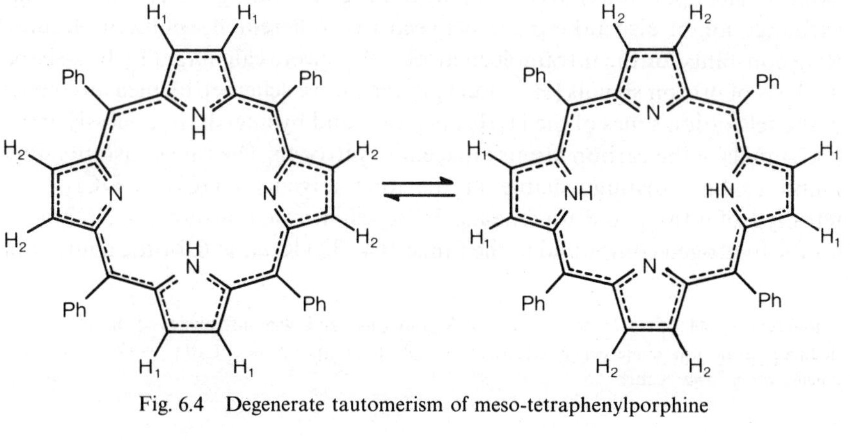

Fig. 6.4 Degenerate tautomerism of meso-tetraphenylporphine

Fig. 6.4 give distinct signals, which broaden and eventually coalesce, the rate of exchange being substantially decreased if the N–H protons are replaced by deuterons. These effects were observed and rate constants calculated for meso-tetraphenylporphine itself [144] and for four para-substituted compounds [145]. The last paper showed that the rates are little affected by the substituent or the solvent. Similar line-shape studies were made on the $^{13}$C-n.m.r. spectrum of the carbon atoms adjacent to the nitrogen atom [444, 445] and on the $^{15}$N-n.m.r. spectrum. Several of these investigations gave abnormally high values for the isotope effect: thus [445] gives $k^H/k^D = 15$ at 298 K, and values of 19, 32 and 48 are given in [145] for 298 K, 263 K and 243 K respectively.* These authors attributed the high isotope effects observed to the simultaneous movement of two protons or deuterons, and the possibility of a tunnel effect was either ignored or was rejected on the grounds that no deviations from the Arrhenius equation or substantial isotopic differences in the pre-exponential factors could be detected. However, closer consideration shows ([447]; cf. Section 4.4.2) that there is no reason to expect abnormally large isotope effects when several atoms move simultaneously, and in none of the investigations so far mentioned is the accuracy or temperature range sufficient to provide definitive evidence about the Arrhenius behaviour. The results in [145] do in fact lead to $E_A^D - E_A^H = 10 \text{ kJ mol}^{-1}$, $A_A^D/A_A^H = 2.3$, both indicative of tunnelling, but the accuracy of these values is not sufficient to warrant any firm conclusions.

A recent very thorough investigation by Hennig and Limbach [448] demonstrates conclusively the role of tunnelling in this system. Magnetic resonance spectra were measured for the nuclei $^1$H and $^{13}$C, and proton resonance was also studied for material containing $^{15}$N in place of $^{14}$N. In the last case the fine structure of the spectra showed that the observed exchange of the inner protons is not due to any intermolecular process but must be purely intramolecular. This demonstration is necessary, since in the presence of traces of acid exchange can take place through the protonated base [449], and the association of porphines in solution by hydrogen bonding [450] suggests that exchange might also take place between two different porphine molecules. Rate constants for the intramolecular exchange were calculated by line-shape analysis of proton signals ($H_1$ and $H_2$ in the above scheme), by measurements of the relaxation times of the $H_1/H_2$ protons, and by line-shape analysis of the $^{13}$C signals of the carbon atoms adjacent to nitrogen. The rate constants were found to be indistinguishable in the four solvents $CDCl_3$, $CDCl_3/CS_2$, tetrahydrofuran-$d_8$ and toluene-$d_8$. Thirty-eight rate constants are reported for the hydrogen compound in the range 164–323 K, and 16 for the deuterium

---

* Reference [144] gives $k^H/k^D = 67$ at 298 K, but this value was not observed directly, being deduced from the coalescence temperatures of the n.m.r. peaks. Later work indicates a considerably lower value.

## Tunnelling in molecular spectra

compound in the range 245–305 K: these are shown in Fig. 6.5 in the form of an Arrhenius plot.

Fig. 6.5 shows very marked deviations from the Arrhenius equation for the hydrogen compound, for which the rate becomes effectively independent of

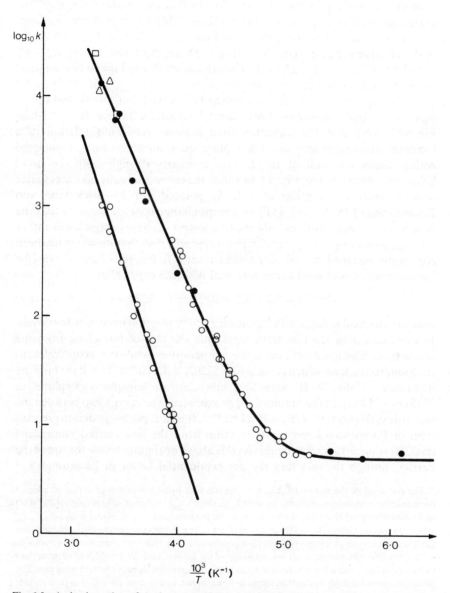

Fig. 6.5 Arrhenius plots for the tautomerism of meso-tetraphenylporphine and meso-tetraphenylporphine-$d_2$ (Reproduced by permission from [448])

temperature below about 200 K. No such deviations are detectable for the deuterium compound, though for this species the rates became too slow to measure below about 240 K. The observed isotope effect $k^H/k^D$ varies from about 12 at 305 K to about 34 at 245 K. These values are high enough to suggest a considerable tunnel correction in this temperature range (cf. the maximum semi-classical values for N–H in Table 4.2), but clearer evidence comes from the slopes and intercepts of the two Arrhenius plots, which are both effectively linear over this range. These yield the values $E_A^D - E_A^H = 14\,\text{kJ mol}^{-1}$, $A_A^D/A_A^H = 25$, both of which are much too high to be explained without a substantial tunnel correction.

All this evidence shows that tunnelling is involved, but this system differs significantly in its quantitative behaviour from any of those so far described. Fig. 6.5 shows that the transition from a linear Arrhenius relation to a temperature-independent rate takes place quite abruptly, being completed within about one unit of $10^3/T$. This contrasts strongly with the usual behaviour, illustrated by Fig. 5.1, in which the curve gradually becomes flatter over a much wider range of $1/T$. As pointed out by Brickmann and Zimmermann [75, 76, 412, 451] an abrupt change of slope suggests that the process concerned involves only the two lowest discrete energy levels rather than the continuous (or nearly continuous) energy distribution which has been commonly assumed in analysing kinetic results. In fact the observed rates for the hydrogen compound agree very well with the expression

$$k^H(s^{-1}) = 5.0 + 3.2 \times 10^{11} \exp(-5220/T) \qquad (6.8)$$

which is identical in form with Equation 3.55 for reaction from only two levels. In this expression the first term represents the rate of tunnelling from the lowest level, which is dominant at low temperatures, while the second refers to reaction from a level which is higher by $5220R = 43.4\,\text{kJ mol}^{-1}$. If we take the frequency of the N–H stretching vibration in tetraphenylporphine as $3315\,\text{cm}^{-1}$ [452], in the harmonic approximation the energy gap between the two lowest vibrational levels is $40\,\text{kJ mol}^{-1}$. It is thus probable that the second term in Equation 6.8 represents reaction from the first excited vibrational level.* It is uncertain whether this level is above or slightly below the top of the barrier, though the fact that the pre-exponential factor in Equation 6.8 is

---

* The geometry of the system (cf. Fig. 6.4) suggests that hydrogen exchange would be effected more readily by bending modes than by stretching, since X-ray studies of solid porphine [453, 454] and tetraphenylporphine [455, 456] show that the porphine skeleton is planar with almost $D_{4h}$ symmetry and that the NH——HN grouping is nearly linear: the same conclusion is reached from the Raman spectrum of tetraphenylporphine in solution [457]. However, there would certainly be several bending vibrational levels below the top of the barrier, and this would lead to more than two terms in Equation 6.8 and hence a more gradual change in the slope of the Arrhenius plot. The participation of bending modes thus seems to be excluded. In any case, the reaction path cannot represent a separable co-ordinate over the whole of its length, and the problems associated with non-separability, discussed in Section 5.3.1, still remain.

considerably smaller than the vibration frequency of $10^{14}\,s^{-1}$ is in favour of the second possibility. The observed activation energy for the deuterium compound is $57\,kJ\,mol^{-1}$, which is close to $2h\nu_{ND}$, suggesting that reaction is proceeding from the second vibrational level. No temperature-independent term could be detected for this compound, and relaxation measurements at 213 K showed that it was less than $0.3\,s^{-1}$.

The first term of Equation 6.8 should persist down to absolute zero, and it is at first sight surprising that at 4.2 K porphine exists in a $n$-octane matrix as two tautomers with NH——HN axes at right angles, which do not interconvert at an observable rate [458, 459]. However, the fact that the two tautomers are distinguishable by their fluorescence spectra and by e.s.r. shows that the symmetry of the molecule is distorted by the forces in the crystal. We are therefore no longer dealing with a symmetrical double minimum potential, and it has already been seen (Section 3.6.3) that even a slight deviation from symmetry greatly reduces the tunnelling frequency. The unique properties of systems with symmetrical double minima are also illustrated by comparing the exchange rate for tetraphenylporphine in solution with those for the isomerization of aryl radicals, described in Section 5.2, which correspond to an unsymmetrical situation. The barrier heights for the two processes are similar, but the distance through which the hydrogen atoms move, i.e. the barrier width, is larger for exchange in tetraphenylporphine (about 180 pm) than for the radical isomerizations (about 130 pm). Nevertheless, the limiting rate at low temperatures is much greater in the former ($5\,s^{-1}$) than in the latter ($10^{-4}$–$10^{-2}\,s^{-1}$). This must be attributed to the special effect of symmetry, which is also responsible for the two-fold difference in the exponents of Equations 2.100 and 2.120. It should be possible to use the observed temperature-independent rate for tetraphenylporphine to obtain information about the nature of the potential energy surface.

It is interesting to speculate on why the results for tetraphenylporphine are best interpreted in terms of reaction from the lowest energy level plus one excited state, whereas the behaviour of the hydrogen atom transfers described in Section 5.2 accords well with the assumption of a continuous energy distribution,* although vibrational modes must be involved in the latter class of reactions also. It appears that these vibrational levels must often be considerably broadened by coupling with other modes of the molecule or its surroundings, thus simulating a continuous energy distribution. Such coupling seems to be unimportant in the porphine system, perhaps because of the rigidity of the molecular framework, the bond lengths of which change by less than 3 pm when migration of hydrogen takes place. This problem is considered further in Section 7.2.3.

* Distinction between these two assumptions is possible only if experimental results are available for rates or isotope effects over a wide range of temperatures, including low temperatures. As mentioned in Section 3.6.3, for small or moderate deviations from classical behaviour the two hypotheses lead to the same qualitative conclusions.

Tunnelling has been invoked to explain features of the $^{19}$F-n.m.r. spectra of some compounds of the type $XF_n$, though here the evidence is much less conclusive. Reviews of this general field have been given by Berry [67] and by Muetterties [460]. Many of these molecules contain fluorine atoms which are in different geometrical situations. For example, $BrF_5$ and $IF_5$ are tetragonal pyramids with four fluorine atoms in one plane and a fifth above the centre of this plane: correspondingly, the $^{19}$F-n.m.r. spectrum has two peaks with areas in the ratio 4:1. $PF_5$ is a trigonal bipyramid with three fluorines in the same plane as the phosphorus and two others above and below this plane. It might therefore be expected to give two $^{19}$F peaks with intensities in the ratio 3:2, but in fact only a single peak is observed [461]. This must mean that the fluorine atoms are changing from one situation to another at a rate which is large compared with the n.m.r. time scale (about $10^{-3}$ s for $^{19}$F). This process could occur by molecular vibrations of large amplitude following approximately one of the normal modes [462], and since fluorine atoms are relatively heavy this would normally be regarded as a classical process with no significant tunnel correction. However, detailed calculations [463] for $PF_5$ suggest that something like one half of the reacting systems have energies below the top of the barrier. Although the estimated tunnelling frequency from the lowest level is low ($\approx 10^{-9}$ s$^{-1}$) the number of vibrational states below the barrier is large, leading to an appreciable tunnel correction. Tunnelling is also favoured by the symmetrical nature of the double minimum potential, as indicated in the last two paragraphs.

Similar problems arise for $ClF_3$ (a bent T) and $SF_4$ (a distorted tetrahedron), both of which contain two kinds of fluorine atoms. In both cases a single $^{19}$F peak is observed at room temperature, changing into a double one as the temperature is lowered [464, 465]. For these molecules, however, appreciable tunnelling seems less likely, since the vibrations concerned also involve motion of the heavier central atom. Finally it should be mentioned that in principle it is also possible for scrambling of fluorine atoms to occur by exchange between molecules. Isotopic exchange of chlorine is known to take place in $PCl_5$ [466, 467], and although such exchange is likely to be much slower for fluorine a direct test is not possible in the absence of any isotopes of this element.

# 7 A review of the present position

## 7.1 Other manifestations of tunnelling

In addition to the topics already discussed there are a number of other physicochemical phenomena which are believed to be significantly affected by the tunnelling of species other than the electron. Three of these will be described briefly here: electrode processes, particularly hydrogen discharge; solid state properties, including the diffusion of hydrogen in metals; and biological implications.

### 7.1.1 *Electrode processes*

Discussion of the role of tunnelling in electrode processes has centred round the cathodic discharge of hydrogen, though similar considerations will apply to the anodic dissolution of hydrogen and possibly to electrode processes involving other light elements.* In the cathodic discharge of hydrogen from an acid aqueous solution the overall process can be divided into several stages, as follows:

(1) The transfer of $H_3O^+$ ions from solution up to the electrode surface.

(2) The actual discharge process, producing hydrogen atoms adsorbed on the electrode surface, i.e.

$$M + H_3O^+ + e^- \rightarrow M-H + H_2O \tag{7.1}$$

(3) The formation (and subsequent evolution) of hydrogen molecules from hydrogen atoms, which can take place by either of two mechanisms (a) and (b),

(a) $\qquad M-H + M-H \rightarrow 2M + H_2$
(b) $\qquad M-H + H_3O^+ + e^- \rightarrow M + H_2 + H_2O$

Any one of these stages may become rate-limiting, depending on the electrode material, the applied potential and the composition of the solution. For acid

---

* In all electrode processes the passage of electrons to or from the electrode is essentially a tunnelling process even over quite large distances ($\approx 1$ nm). However, this is not the rate-determining process in hydrogen discharge, and the tunnelling considered in the present context refers to the motion of protons or hydrogen atoms.

solutions and metals of high overpotential, notably mercury, the rate-determining stage is usually the second, represented by Equation 7.1, and since this involves the motion of a proton across a potential barrier the possibility of considerable tunnel corrections arises. This was first clearly formulated by Bawn and Ogden [41] in 1934, and more detailed treatments were given subsequently by many authors [66, 468–498].

The problems which arise are similar to those involved in homogeneous proton-transfer reactions, and evidence for tunnelling can be sought in the rates and temperature coefficients of the discharge process, and especially of the isotope effects measured by the H/D and H/T separation factors. There is also the additional factor that the height of the energy barrier can be varied continuously over a large range by varying the potential applied to the electrode: in a homogeneous proton transfer this can only be achieved in a discontinuous manner by chemical substitution in the reactants. The effect of the potential is usually expressed by the Tafel equation

$$\eta = a + b \log i \qquad (7.2)$$

where $\eta$ is the overpotential, $i$ the current density, and $a$ and $b$ are effectively constant over a limited range. The interpretation of the experimental results is often based on the effect of temperature or isotopic substitution on the values of $a$ and $b$.

Most (though not all) investigators have concluded that considerable tunnel corrections are involved. Particular attention has been paid to the effect of the cathode potential on the isotopic separation factor: for example, for a mercury cathode in solutions of strong acids the H/T separation factor varies from 3.4 to 9 as the cathode potential changes from $-0.7\,\text{V}$ to $-1.3\,\text{V}$. This could be formally accounted for in terms of the 'symmetrical' vibrations of the transition state [along the lines of Westheimer's explanation of reduced isotope effects in homogeneous proton transfers (cf. Section 4.2.4)], but it appears that this explanation is inadequate to account for the observed facts, and a tunnelling interpretation is preferred. However, no unanimity has been reached about the details of the tunnelling calculation, as evidenced by the polemical nature of several recent publications [499–503]. The points at issue include the methods of estimating the barrier dimensions, the importance of the electrical double layer, the discrete or continuous distribution of energy levels in $H_3O^+$, and the part played by solvent reorganization. The last two points reflect a general dichotomy in current views about proton-transfer processes in polar media, and will be discussed in Sections 7.2.3 and 7.2.4.

7.1.2 *Solid state phenomena*

Tunnelling has been invoked to account for a wide variety of phenomena in solids, especially at low temperatures. This arises particularly in the effect of tunnelling on the rotation of groups, molecules and ions, and the bearing of this on magnetic resonance spectra has already been described in Section 6.4. Rotational tunnelling also affects a number of other optical, thermal and mechanical properties of solids, and the same is true for the tunnelling motion of defects between equivalent sites. For example, if small concentrations of lithium ions are present in a crystal of KCl they can tunnel between several equivalent off-centre positions in the unit cell, and it is interesting to note that the appropriate isotope effect has been observed for $^6$Li and $^7$Li. Similar behaviour is shown by other systems, such as NaBr–F$^-$ and RbCl–Ag$^+$. This general field has been recently reviewed [504, 505]. Apart from crystals the tunnelling motion of atoms or groups has been invoked to explain the low-temperature thermal and visco-elastic properties of glasses and polymers [506, 507]. The tunnelling of hydrogen between two equivalent positions in a hydrogen-bonded solid has already been mentioned in Section 6.3 in connection with infra-red spectra. In the case of the alkali dihydrogen phosphates further evidence comes from low-frequency Raman spectra [508] and from neutron scattering [509], and the ferro-electric Curie point of these substances is also believed to be influenced by the tunnel effect [510].

The rate of diffusion of H, D and T-atoms in metals has been much investigated and shows some peculiar features [214–216, 511], including inverse isotope effects, some of which may be due to the existence of discrete energy levels for the proton in different sites in the crystal [214, 512]. These problems have often been treated in terms of the tunnel effect, but the approaches of different investigators differ fundamentally in several respects. One interpretation [513] treats the motion of the proton by transition state theory and applies a tunnel correction for a parabolic barrier with a continuous energy distribution. A second treatment [216] uses the picture of a sinusoidal potential with tunnelling from several levels at frequencies determined by the splitting of these levels. Yet another view [514] supposes that the activation energy is provided not by proton motion but by movements of the metal lattice: this picture has much in common with the views of the Russian school, discussed in Section 7.2.4, about the role of the solvent in proton transfers in polar media.

### 7.1.3 *Biological implications*
Many biological processes of fundamental importance involve the motion of a proton in a hydrogen bond, i.e. X—H——Y → X——H—Y, and may therefore be considerably influenced by tunnel corrections. This interpretation has been particularly stressed by Löwdin [74, 515–517] in connection with the genetic information contained in DNA: this is associated with hydrogen bonds between base pairs, and any disturbance of this pairing can lead to mutations. Theoretical studies, especially of tautomerism in the guanine–cytosine pair [518–527], suggest that tunnelling should be important in such systems. However, it is difficult to point to any clear experimental con-

sequences of the tunnel effect, and a study of the rates of mutation to phage resistance in $H_2O$ and $D_2O$ did not reveal the large isotope effect which might have been expected [527]. Another biological system in which tunnelling has been postulated is in proton conduction through polypeptides and protein helices, which is important for transmission across membranes and along fibres [528–531]. It again seems uncertain whether this interpretation follows necessarily from the experimental facts.

The discovery that tunnelling processes can proceed at an appreciable rate at extremely low temperatures even when the motion of heavy entities is involved (Section 5.6) may have a bearing on the origin of life in the universe [532]. Interstellar dust particles are believed to contain a mixture of frozen water, methane, ammonia and other molecules containing up to nine atoms [533, 534]. The formation of such molecules can be envisaged in terms of the reactions of ions or radicals formed initially by ultra-violet radiation [535–537], but since radiation can also decompose polyatomic molecules it is difficult on the basis of classical kinetics to see how substances of high molecular weight could be formed at the low temperatures (10–20 K) of interstellar space [538]. Nevertheless, there have been recent reports of the presence in galactic dust of polyoxymethylene [539–541] and polysaccharides [542], and it is of particular interest that studies of the polymerization of formaldehyde to polyoxymethylene have given clear evidence of the persistence of tunnelling processes down to 4 K [362–369]. These ideas open up very interesting possibilities for what has been termed the 'cold pre-history of life', though the subject is obviously still in its infancy.

## 7.2 Current problems in the theory of tunnel corrections

The remainder of this chapter will be devoted to a critical examination of some of the assumptions which are commonly made in the theory of tunnel corrections in chemical kinetics. In spite of the qualitative and at least semi-quantitative success of such theories, as described in Chapter 5, they contain several features which are open to criticism, though feasible alternatives are not often available. The topics treated are the validity of transition state theory, the applicability of one-dimensional tunnel corrections, continuous versus discrete energy distributions, and the role of the solvent in tunnelling. Although these problems interact with one another to some extent it is convenient to consider them separately.

### 7.2.1 *The validity of transition state theory*
Tunnelling problems are most commonly treated by applying some form of tunnel correction to semi-classical transition state theory, so that the validity of the latter must be examined. The fundamental expression, Equation 3.17, has been derived in a number of equivalent ways, which will not be described

here: a general account of the arguments and assumptions has been given recently by Marcus [543]. The necessary assumptions can be summarized as follows:

(a) In the space (hyperspace) represented by the co-ordinates and momenta of the particles involved it is possible to define a dividing surface (hypersurface) such that all reactants crossing it go on to form products.

(b) The transition state, comprising systems crossing the dividing surface in a forward direction, is in statistical equilibrium with the reactants. (This does *not* imply that systems crossing the dividing surface in the reverse direction are in equilibrium with the reactants, unless the whole assembly of systems is in a state of equilibrium.) Another aspect of the assumption is that the number of systems crossing the dividing surface in the forward direction is independent of those crossing in the reverse direction.

(c) A corollary of (b) is that there is internal statistical equilibrium between the degrees of freedom of each type of system, whether it is a reactant, a transition state or a product.

(d) Motion through the transition state can be treated as a classical translation.

Of the above assumptions, the possible failure of (d) is closely bound up with the whole question of tunnel effects, and will be considered in Section 7.2.2.

As regards (a), examination of typical energy surfaces shows that it will usually be possible to choose a dividing surface for which this assumption is a good approximation. This surface will normally be at the highest point of the path of minimum energy connecting the reactants and the products. This means that the classical transmission coefficient $\kappa$ in Equation 3.17 (not to be confused with the quantum-mechanical barrier permeability $G$ or the tunnel correction $Q_t$) is close to unity. There are two situations in which this assumption might be substantially in error. The first arises if the energy surface includes complicated features, such as an energy basin close to the transition state, which can cause an appreciable number of systems to be 'reflected' back to the initial state. The second applies to reactions in which there is a switch from one quantum state to another: the process then involves a transition between two energy surfaces, the probability of which may be considerably less than unity, especially if a change in spin angular momentum or multiplicity occurs. Neither of these situations is likely to arise in the transfers of protons or hydrogen atoms which are the main classes of reaction for which tunnel corrections have been applied. The assumption that $\kappa = 1$ thus appears to be justified in the present context.

Assumptions (b) and (c) can never be strictly correct, since chemical reactions remove molecules which possess energy in excess of the average or which have a particular type of energy distribution. These molecules can only be replenished at a finite rate, and the occurrence of the reaction must therefore cause some perturbation in the equilibrium energy distribution. However, it seems certain that this perturbation will only have a small effect on the reaction velocity under normal conditions. A number of authors [544–548] have attacked the problem theoretically on the basis of simplified models. Although the methods of approach and the models used vary widely the general conclusions are unanimous, namely that the effect of disequilibrium on the reaction velocity is small unless the activation energy is low. Typical figures for this effect are 10–20% when the activation energy is in the range 5–10k$T$. Moreover, most of these calculations apply only to reactions of simple molecules in the gas phase, and it is clear that the equilibrium hypothesis will be favoured by increasing molecular complexity and by carrying out the reaction in a condensed phase, since both of these factors will increase the number of paths for the redistribution of energy. The validity of the equilibrium hypothesis is also supported by experience, in that for a given reaction the same value of the velocity constant is found to represent the experimental results when the system is displaced from equilibrium to various extents. The equilibrium hypothesis of transition state theory thus seems to be justified, except for very fast reactions of simple molecules in the gas phase.*

Attempts have been made to assess the validity of transition state theory by comparing its results with those of a detailed dynamical treatment in which the equations of motion are solved for individual trajectories and the overall reaction velocity obtained by an averaging procedure. Such calculations are feasible only for very simple reactions (in practice mainly those involving three hydrogen atoms) and they have usually been made on the basis of empirical energy surfaces which may differ considerably from the true ones: however, the last point is immaterial in the context of making comparisons between the transition state and trajectory calculations. Comparisons of this kind on a purely classical basis for $H + H_2$ show reasonably good agreement in the range 300–1000 K, though there are some deviations at lower temperatures [550, 551]. Similar calculations [292–294] for $H + Cl_2$ show virtually exact agreement at 450 K and a discrepancy of about 30% at 200 K. Because of the simple nature of the reactions and the absence of a solvent these examples constitute a particularly severe test of transition state theory, which is likely to apply even more closely to the generality of chemical reactions.†

---

* It should be stressed that transition state theory and simple collision theory are equally dependent on an equilibrium hypothesis, since the latter assumes that the collision frequency and the energy distribution are the same as they would be in the absence of reaction. Collision theory points out more clearly the fact that not only the energy distribution but also the spatial distribution of molecules (and hence the collision number) can in principle be perturbed by reaction: pairs of molecules in close proximity are removed and are re-formed by diffusion processes of finite velocity. However, a quantitative treatment shows that the effect of diffusion becomes significant only for very fast reactions or in media of high viscosity [549].

† Comparisons have also been made between the results of transition state theory and of various quantum-mechanical treatments of simple reactions [273–279], but since the latter automatically include an allowance for tunnelling these comparisons are not relevant to the present section.

# A review of the present position

The general conclusion is therefore that, in the absence of any appreciable tunnel correction, transition state theory is a good approximation to the truth for most problems of chemical kinetics. In any case, its continued use in chemistry can be justified by the absence of any feasible alternative, except perhaps for a few very simple reactions in the gas phase. The attitude of the average chemist is well expressed by the moral of Hilaire Belloc's cautionary tale [552] about the little boy who was eaten by a lion in the zoo, namely

> And always keep a-hold of Nurse
> For fear of finding something worse.

### 7.2.2 The applicability of one-dimensional tunnel corrections

It is easily seen that assumption (d) of the last section, the treatment by classical mechanics of motion along the reaction co-ordinate, is not really justified for the movement of light nuclei, especially the isotopes of hydrogen. In the commonest version of transition state theory the transition state is defined so as to include a small length $\delta$ of the reaction co-ordinate in the neighbourhood of the dividing surface. If $\delta$ (which cancels out in the final expression) is made sufficiently small the potential energy along the reaction co-ordinate in the transition state is effectively constant, and motion along this co-ordinate can be treated as a translation. From this point of view the value of $\delta$ ought not to exceed 10 pm at the most. On the other hand, according to the uncertainty principle the motion of a particle can be described classically only over distances which are considerably larger than the de Broglie wavelength. Since this wavelength is 100 pm for protons moving with thermal velocities at 300 K these two requirements are clearly incompatible, and this incompatibility is another way of expressing the need for a tunnel correction in reactions involving the motion of light particles.

The usual procedure, adopted throughout this book, is to treat motion along the reaction co-ordinate as a one-dimensional barrier problem for which a tunnel correction factor can be computed, and then to multiply the semi-classical expressions of transition state theory by this factor. As already pointed out in Section 5.3.1 this procedure is justified only to the extent that the reaction co-ordinate is *separable* in the mathematical sense, which means that in the overall expression for the kinetic and potential energies this co-ordinate must only occur in terms which do not involve any of the other co-ordinates describing the system. When Cartesian co-ordinates are being used such separation occurs only when the potential energy along the reaction co-ordinate is a quadratic function of the displacement, i.e. when the barrier is parabolic*. Examination of theoretical surfaces for simple systems shows that

---

* The same situation arises in the vibrational analysis of a stable species, when separation into normal modes is possible only when the vibrations are simple harmonic. The position is modified somewhat if curvilinear co-ordinates are used in place of Cartesians [553, 554], but the problem of separability still remains.

the barrier is approximately parabolic over a distance of only about 30 pm, so that a separate treatment of the tunnel correction is legitimate only when this correction is small. Although the profile along the reaction path can be represented much more realistically by non-parabolic barriers such as the Eckart or Gaussian functions (Equations 2.45 and 2.68) there is no strict justification for applying one-dimensional tunnel corrections based upon such barriers.

Even if the problem of non-separability is ignored there is another difficulty in the method of treatment usually adopted for a reaction of the type $A-H + B \to A + H - B$, where A and B are heavy groups. In the usual method of plotting the energy surface for such a system, shown in Fig. 3.2, a motion in which H moves while A and B are almost stationary is represented by a line of slope $-45°$. This condition is approximately satisfied for the central part of the reaction path, but not for its initial and final portions, which correspond to considerable movements of A and B. A logical treatment of tunnelling through a 'straightened out' reaction path should therefore employ a variable reduced mass, equal to $\frac{1}{2}m_H$ near the top of the barrier (cf. Equation 3.11), but rising to much larger values as the initial or final state is approached. Although it is possible to carry out such a treatment it has been more usual to assume that the reduced mass remains constant along the whole path, which will clearly exaggerate the role of tunnelling.

There is no simple answer to the dilemmas raised in the last two paragraphs. A stricter treatment would allow for tunnelling in more than one dimension, i.e. it would consider all possible tunnelling paths connecting the initial and final valleys in a diagram such as Fig. 3.2. Such calculations are intractable and do not automatically circumvent the problems of non-separability. The term 'tunnelling' acquires an extended meaning when more than one dimension is being considered, since it includes not only reaction paths which are classically forbidden because the energy of the system is too low, but also those for which the energy is higher than the saddle point in the energy surface, but which are dynamically forbidden in classical mechanics because the lowest point of the path lies still further above this saddle point [555, 556]. This distinction does not arise in tunnelling in one dimension, for which the energy criterion is sufficient.

Because of these difficulties there have been various attempts to circumvent the shortcomings of the usual one-dimensional treatment by more or less empirical procedures, some of which have already been mentioned in Section 5.3.1 in connection with the H——H——H reaction. The most commonly used procedure is that proposed by Johnston and Rapp [89] for reactions of the type

$$A - H + B \to A + H - B.$$

This employs a number of cuts at $-45°$ through the energy surface, as

## A review of the present position

indicated by the broken straight lines in Fig. 5.2. The energy profile along each cut is fitted to an Eckart barrier, and the tunnel correction calculated on the basis of a fixed effective mass ($\frac{1}{2}m_H$ if A and B are heavy groups or atoms). The effective tunnel correction is then calculated by averaging over the different paths: almost the same value is obtained for the most probable $-45°$ path, the apex of which has an energy and interatomic distances somewhat greater than the values at the saddle point. This procedure was applied to the H——H——H reaction, using some of the semi-empirical energy surfaces referred to in Section 5.3.1.* The tunnel correction calculated by the procedure of Johnston and Parr is considerably smaller than that obtained by a one-dimensional treatment of the 'straightened out' reaction path: at 333 K the values are 6 and 16 respectively. However, the system of three hydrogen atoms certainly represents an extreme case, and rather different general conclusions have been reached recently by Christov and Georgiev [557]. These authors used the semi-empirical energy surface obtained for H——H——H by the method of Sato [289], but assumed that the two end atoms were of infinite mass: they also replaced the Eckart barrier by a modified version [66] which provides a much better fit to the energy surface. When the central atom is the light isotope of hydrogen this model gives a tunnel correction at 300 K of 4.9 by the Johnston–Parr procedure and 6.7 for a one-dimensional 'straightened out' barrier. The difference between these two values is certainly less than the uncertainty due to our ignorance of the true energy surface for any specific reaction. This suggests that the one-dimensional treatment is a reasonable method of interpreting experimental results when, as is usually the case, there is no firm independent basis on which to construct a barrier or a surface.

There have been other recent suggestions for improving the calculation of one-dimensional tunnel corrections. One of these [282] is equivalent to averaging the tunnelling probability for a time-dependent one-dimensional barrier, and thus bears some resemblance to the Johnston–Rapp procedure, in which the averaging occurs over a number of one-dimensional barriers obtained as sections of the energy surface. This treatment does not appear to have been compared with experiment or with the results of more exact theoretical calculations. Another suggestion [272] is that the 'straightened out' barrier should be taken not along the path of lowest potential energy, but along the limit of classical vibration amplitudes, as indicated by the broken curve in Fig. 5.2. This makes the effective barrier higher and narrower, and over part of its length it will be close to the most probable $-45°$ cut in the Johnston–Rapp treatment. There is no clear theoretical basis for this procedure, and it is difficult to see how it avoids the problem of non-separability, but it is stated to improve agreement with quantum-theoretical calculations.

These suggestions represent aspects of various attempts to formulate general

---

* As pointed out by Christov and Georgiev [557], the argument for restricting cuts to an angle of $-45°$ is a strong one only when we are considering the motion of a light particle between two heavy centres, as may be seen from Equation 3.13 for the efffective mass. The application of this procedure to reactions of three hydrogen atoms is therefore open to question.

modifications of conventional transition state theory which will include non-classical motion along the reaction path and at the same time avoid difficulties arising from non-separability. This approach has been followed particularly by Miller [90, 280, 281]: the treatment appears to owe a good deal to intuition, and it is necessary to use semi-classical (BWK) approximations in solving the Schrödinger equation. Good agreement with experiment is claimed for the reaction $D + H_2$ at 200 K and 1000 K, using the best energy surface available [269, 270]. The amount of computation involved is considerable even for this very simple reaction, and since reliable energy surfaces are not available for more complex systems further practical applications are at present remote.

In any reaction involving considerable departures from classical behaviour a counsel of perfection would be to use exact quantum theory to calculate the reaction probabilities for all possible initial rotational and vibrational states and all possible trajectories, and then to obtain the reaction velocity by an averaging procedure. In such a computation the tunnel effect does not appear as a separate issue, but should be automatically allowed for in calculating the quantum-mechanical transition probabilities. It is of course necessary to know the energy as a function of the nuclear positions (i.e. the energy surface or hypersurface), and although the principles of the methods are well understood the task of computation is a severe one: thus recent calculations for the reaction $H + H_2$ over a range of temperatures required about four hours of computer time on a large computer [273]. It is therefore quite unrealistic to contemplate the application of this method to more complex systems or to reactions in condensed phases.

There is thus considerable justification for continuing to interpret experimental results in terms of one-dimensional tunnel corrections, even though the form of the barrier derived may not correspond exactly to any feature of the true energy surface. Moreover, there are some processes which can be represented, at least to a very good approximation, by the motion of a particle along a single co-ordinate.* This is obviously the case for hindered internal rotation, in which the angle of rotation determines the potential energy. It is also a good approximation for processes in which a proton or a hydrogen atom is transferred between heavy centres either intramolecularly or in the solid state. As discussed in Section 3.2, the potential energy of the system then depends on only a single co-ordinate and the reduced mass to be used in the barrier problem is approximately equal to the mass of the particle transferred rather than the smaller mass represented by Equation 3.13.

### 7.2.3 Continuous versus discrete energy distributions

Most applications of tunnel corrections have assumed a continuous Boltzmann distribution of energies, as expressed by Equation 3.1. This assumption has been retained in the whole of Chapter 3, with the exception of Section 3.6.3, and it is the basis of most of the quantitative interpretation of

---

* This need not be a Cartesian co-ordinate, but may alternatively be an angle or some other curvilinear co-ordinate.

# A review of the present position

experimental results described in Chapter 5. However, the energy of molecular vibrations is certainly quantized, and for stretching vibrations of X–L (where L is an isotope of hydrogen) the separation of the vibrational levels is much greater than $kT$ for the range of temperatures normally encountered. It might appear, therefore, that tunnelling from one or more discrete levels is a more appropriate model than that of a continuous energy distribution, especially at low temperatures. As shown in Section 3.6.3, the two pictures lead to results which differ quantitatively, though they are qualitatively similar.

In a bimolecular gas reaction the assumption of a continuous energy distribution is fully justified, since the translational kinetic energy of the reacting species is not quantized. This conclusion can be reasonably extended to bimolecular reactions in solution, since although the solute molecules can be pictured as oscillating in a solvent cage these oscillations will have low frequencies and the closeness of their energy levels will simulate a continuous distribution. In addition to this the solvent cages will not all be identical and will fluctuate with time, thus producing a further 'smearing out' of the energy levels. The picture is less appropriate when some of the reactant or product species are charged and the process is occurring in a polar medium, as is the case for many proton-transfer reactions. There are then strong solute–solvent interactions which vary during the reaction process, and although these interactions will lead to a quasi-continuous distribution of energies for the system as a whole this does not necessarily imply that such a distribution is appropriate for calculating tunnel corrections. The whole question of the role of the solvent in tunnelling will be considered in Section 7.2.4.

In intramolecular transfers of protons or hydrogen atoms there is no contribution from translational kinetic energy and the assumption of a continuous energy distribution appears less reasonable. The simplest picture of such transfers is that they involve excitation of X–H stretching vibrations, and in this event reaction at ordinary temperatures might be expected to involve only two levels, one near the bottom of the barrier and the other close to or above the top. Experimental evidence on this point is conflicting, as already described in Section 5.2. The extensive measurements of Ingold and his collaborators on the isomerization of hindered aryl radicals [231–233] agree well with predictions based on a continuous energy distribution and definitely cannot be represented by a two-level model, which, as shown in Section 3.6.3, predicts a fairly sharp change of apparent activation energy over a limited temperature range.* The success of the treatment in terms of a continuous energy distribution suggests that the vibrational levels for C–H stretching are effectively broadened to a considerable extent by the super-

---

* The results could presumably be fitted within experimental error by assuming the participation of three or more energy levels, but this involves the introduction of more adjustable parameters and rapidly becomes indistinguishable from a continuous distribution.

position of other vibrational levels of much lower frequency, for example C–H bending or vibrations of the aromatic skeleton.* Alternatively, the presence of flexible groups in the radicals investigated may lead to the existence of a range of barriers with different parameters: since the observed reaction velocity represents an average over all these situations the effect is equivalent to a broadening of the energy levels. It may be noted that the observed rate at the lowest temperature investigated (28 K) is somewhat higher than the value predicted from the Eckart or Gaussian barriers fitted to the results between 77 K and 233 K. This may be due to the breakdown of the assumption of a continuous energy distribution, since even vibrational modes of very low frequency will be almost exclusively in their lowest states at temperatures below about 40 K, so that no further decrease in reaction velocity would be expected below that temperature.

Similar considerations apply to intermolecular hydrogen atom transfers in solids and glasses, especially the reactions of the type

$$CH_3^{\cdot} + CH_3X \rightarrow CH_4 + {}^{\cdot}CH_2X$$

studied by Ffrancon Williams and his collaborators ([219–224]: cf. Section 5.2). Since the reactants are immobilized in the solid or glass the transfer of hydrogen is likely to involve the C–H stretching vibration: nevertheless, the results can be interpreted on the basis of a continuous energy distribution, and could not be accounted for by a two-level model. This can be attributed to the effect of the lattice vibrations of the solid (phonons), either in smearing out the molecular energy levels or by providing a range of barriers of different dimensions. The lowest temperature range investigated was 10–90 K, for the reaction of methyl radicals with methanol [222, 223]. The rate of this reaction becomes effectively independent of temperature below 40 K, though the barrier derived from the results at 40–90 K predicts a continued decrease with decreasing temperature if a continuous energy distribution is assumed. The observed discrepancy may therefore again be due to the failure of this assumption, though the authors [223] have given an alternative explanation in terms of the freezing of rotation of the methyl group at temperatures below 40 K.

An analogous situation obtains in the transfer of hydrogen atoms from

---

* The idea of broadened energy levels in a kinetic context does not conflict with the observation of narrow bands in vibrational spectra. Theory shows that these bands correspond to normal vibrations of the species concerned: bands corresponding to the sum of more than one normal frequency are much weaker and are often forbidden by the selection rules. Nevertheless, a molecule can be thermally excited simultaneously in several normal modes, which may combine to bring about reaction in a way which is forbidden in interaction with radiation. (In other words, motion along the reaction co-ordinate need not coincide even approximately with any normal mode.) A simple example is provided by translational energy, which can certainly contribute to bimolecular reaction processes, but has no effect upon spectra.

dimethylglyoxime molecules to iminoxy-radicals in $\gamma$-irradiated crystals of dimethylglyoxime [235]: cf. Section 5.2). For the temperature range 50–145 K the results can be accounted for by tunnelling through an Eckart barrier with a continuous energy distribution. An isolated value at 4.2 K is almost the same as at 50 K, and about ten times that predicted by the above treatment, again suggesting that the assumption of a continuous energy distribution becomes invalid at these very low temperatures.

The position is different for intramolecular hydrogen migration in meso-tetraphenylporphine, which has been thoroughly investigated by nuclear magnetic resonance between 164 K and 323 K ([448]:cf. Section 6.5 and Fig. 6.5). For this system there is quite an abrupt change in the apparent activation energy, which falls from 43 kJ mol$^{-1}$ to zero within about one unit of $10^3/T$. This strongly suggests that only two energy levels are involved, and the results do in fact agree quantitatively with this hypothesis, as is shown by Equations 3.55 and 6.8. Moreover, the value of about 40 kJ mol$^{-1}$ deduced for the energy difference between the two levels indicates that only N–H stretching vibrations are involved, since if bending vibrations also contributed there would be several levels below the top of the barrier and a more gradual change in the slope of the Arrhenius plot should be observed. (This conclusion is somewhat surprising, since the geometry of the system suggests that bending motions should be more effective in transferring hydrogen atoms.) The behaviour of this system thus differs considerably from that observed for intramolecular hydrogen transfer in radicals and intermolecular hydrogen transfer in solids, described in the preceding paragraphs. The special position of the tetraphenylporphine system may be a consequence of the rigidity of the porphine skeleton, which makes the participation of skeletal vibrations less likely, or it may depend on the fact that in this system the double-minimum potential is truly symmetrical, which is not the case for the other hydrogen-transfer reactions.

The general conclusion of this section is that in tunnelling calculations the assumption of a continuous Boltzmann energy distribution usually works much better than might be expected. It must necessarily fail at very low temperatures, and there is some experimental evidence to support this. However, in most situations is seems justifiable to retain the assumption of a continuous energy distribution, since information is rarely available about the location of discrete levels in relation to the barrier, or the extent to which they are broadened. In any case it should be emphasized that, as stated in Section 3.6.3, the two pictures lead qualitatively to the same results, so that it is usually difficult to distinguish between them experimentally.

### 7.2.4 The role of the solvent in tunnelling

Although much of the evidence for significant tunnel corrections derives from observations on processes in solution, most treatments of the problem pay scant attention to any involvement of the solvent. This is a reasonable point of view for hydrogen atom transfers, since these are homolytic processes not involving any appreciable charge separation, so that solute-solvent interaction is not likely to change significantly in the sequence: reactants → transition state → products. This view is supported by experimental findings on the effect of varying the solvent for several processes of this type [231–233, 145, 448]. Although these systems reveal striking departures from classical behaviour neither the rates nor the isotope effects are appreciably affected by considerable changes in the nature of the solvent, or, in some instances, by transferring the process to a solid matrix. It is true that these comparisons have only been made for intramolecular hydrogen transfers, which should be least sensitive to changes in the environment, and it would be desirable to have similar information for intermolecular reactions. However, it seems likely that the general insensitivity to solvent changes of the rates of bimolecular homolytic processes will extend to instances in which there are significant tunnel corrections.

The position is different in reactions involving the transfer of a proton rather than a hydrogen atom, many of which show features characteristic of appreciable tunnel corrections. In these protolytic or acid–base reactions some of the species involved must be charged, and will therefore interact strongly with solvent molecules, especially if these are polar. The occurrence of a protolytic reaction will thus normally be accompanied by changes in solvation, or solvent re-organization. The importance of solvation changes in determining acid–base equilibria is well established and at least qualitatively understood [558]. On the other hand, quite apart from any consideration of tunnel corrections, there is still much controversy regarding the part played by solvent re-organization in the kinetics of proton-transfer processes, in particular the extent to which such re-organization has occurred in the transition state, and whether it precedes, follows, or is synchronous with the actual transfer of the proton.

A commonly used reaction scheme is based largely on the work of Marcus [197, 559]. The assumed sequence of events is shown in Equations 7.3: these have been written for the reaction

$$AH + B \rightleftharpoons A^- + HB^+$$

but apply to any choice of charges on the species AH and B.

$$
\begin{aligned}
&\text{(a)} \quad AH + B \rightleftharpoons AH \| B \\
&\text{(b)} \quad AH \| B \rightleftharpoons AH \cdot B \\
&\text{(c)} \quad AH \cdot B \rightleftharpoons A^- \cdot HB^+ \\
&\text{(d)} \quad A^- \cdot HB^+ \rightleftharpoons A^- \| HB^+ \\
&\text{(e)} \quad A^- \| HB^+ \rightleftharpoons A^- + HB^+
\end{aligned}
\quad (7.3)
$$

Considering each process in the direction from left to right, step (a) represents the diffusion together of AH and B to form the *encounter complex* AH||B, in which the reactants have come together, but do not interact appreciably and have not undergone any significant modification of their solvation shells. Step (b) represents the conversion of the encounter complex into the *reaction complex* AH·B, in which the reactants and the neighbouring solvent molecules are in positions and orientations favourable to proton transfer. This step may involve rotation of the reactants, solvent re-organization, and also in some instances a re-structuring of the heavy-atom framework of A and B. Step (c) is the actual proton-transfer process, which is supposed to occur without appreciable movement of solvent molecules or heavy atoms. Finally, step (d) represents the conversion of the reaction complex $A^-\cdot HB^+$ into the encounter complex $A^-||HB^+$, again involving solvent re-organization and perhaps heavy atom movement, and (e) is the diffusion apart of $A^-$ and $HB^+$. The sequence of events for the reverse reaction is of course entirely analogous.

If the above scheme is accepted, the observed rate (or free energy of activation) of the overall forward reaction contains contributions from each of the steps (a), (b) and (c). By analysing the dependence of the reaction velocity on acid–base strength (i.e. on the equilibrium constant of the reaction) for a number of systems several authors [106, 197–201, 560, 561] have concluded that step (b), involving solvent re-organization, contributes a considerable proportion of the measured free energy of activation. Although the numerical values obtained must be accepted with reserve, since they depend on a number of simplifying assumptions, this conclusion appears to be qualitatively correct within the framework of the scheme in Equation 7.3. Since the tunnel effect can only be important for the actual proton-transfer step (c), this implies that any quantitative treatment of tunnel corrections refers only to this step, and not to the reaction as a whole. In particular, the barrier height in the tunnelling problem need not be directly related to the observed activation energy but may be considerably less than this quantity, at least for proton transfers in polar media such as water. This obviously affects the interpretation of barrier dimensions derived from experimental results, such as those given in Table 5.5. This is often unimportant, since frequently only the approximate magnitude of these dimensions is of interest, but the part played by solvent re-organization would certainly have to be taken into account if calculations of tunnelling are to be compared quantitatively with molecular models, a prospect which at present seems remote as far as proton-transfer reactions are concerned.

There is a second and perhaps more important way in which interaction with the solvent may impinge upon the operation of the tunnel effect. In the treatment described in the last two paragraphs it is assumed that there is no coupling between the motion of the proton and the motion of the solvent or of other heavy atoms. This assumption is equivalent to the Franck–Condon principle, and is taken over from the theory of electron-transfer processes in solution. However, although electron-transfer and proton-transfer processes present formal similarities there are important quantitative differences between them. The electron is 2000 to 200 000 times lighter than the nuclei with which it is associated, while the corresponding factor for the proton is usually only 10 to 20.* It therefore seems intuitively likely that the motion of the proton will be accompanied (and not merely preceded or followed) by some movement of solvent molecules, in particular by rotation of dipolar molecules such as water. The effect of this coupling will be to increase the effective mass for tunnelling above that of the proton, and hence to decrease the tunnel correction, which is sensitive to the mass of the tunnelling particle. The magnitude of the tunnel correction, and hence of observable properties such as the isotope effect, may then depend considerably on the nature of the solvent, and in particular on its polar characteristics.

The way in which the increased effective mass arises may be illustrated by considering a simple electrostatic model, shown in Fig. 7.1. The proton is moving along the $x$-axis and a dipolar solvent molecule of moment of inertia $I$ is at a distance $l$ from the proton, with its negative end pointing towards the proton and its axis making an angle $\theta$ with the $x$-axis. It is assumed that when the proton moves through a small distance $\Delta x$ the dipole rotates through an angle $\Delta\theta$ so that it is still pointing towards the proton. Simple geometry then gives $l\Delta\theta = \Delta x \sin\theta$, and the combined kinetic energy of the translating proton and the rotating solvent molecule is

$$T = \tfrac{1}{2}m_H \dot{x}^2 + \tfrac{1}{2}I\dot{\theta}^2 = \tfrac{1}{2}\dot{x}^2[m_H + (I\sin^2\theta/l^2)] \tag{7.4}$$

giving for $m'_H$, the effective mass for proton transfer

$$m'_H = m_H + (I\sin^2\theta/l^2). \tag{7.5}$$

This expression gives an upper limit for the increase in effective mass, since it assumes that there is no lag between the motion of the proton and the rotation of the solvent molecule. This lag will be least when the solvent has a large dipole moment, a quantity which does not occur in the above expressions.

A similar increase in effective mass and decrease in tunnelling is predicted by a more general model [209], in which the coupling with the solvent is represented by a bending force constant between the solvent dipole and the proton. This last treatment produces a rather surprising prediction for the effect of coupling between the motion of the proton and the motion of heavy atoms in the reactants, a situation which must often arise when the abstraction of a proton to give a carbanion is accompanied by structural

---

* The justification for the separate treatment of the movement of light and heavy particles, as embodied in the Franck–Condon or Born–Oppenheimer principles, involves an expansion in powers of $(m/M)^{1/4}$ and subsequent neglect of higher terms, $m$ and $M$ being respectively the masses of the light and heavy particles. For electron transfer $(m/M)^{1/4} < 0.16$, which makes the approximation a reasonable one, but for proton transfer the corresponding inequality is $(m/M)^{1/4} < 0.6$, which hardly justifies the analogy with electron transfer. The same point is elaborated later in this section in connection with the theories of Dogonadze and his collaborators.

# A review of the present position

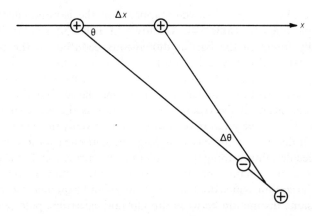

Fig. 7.1 Model for rotation of solvent dipole in field of moving proton

re-arrangement, as for example in the ionization of ketones or nitro-alkanes. It is found that the effect of such coupling is to increase the tunnel correction, since although there is an increase in the effective mass this is more than compensated for by a more sharply curved barrier.

A natural experimental approach to the question of solvent participation in tunnelling would be to investigate the effect of varying the solvent on tunnel corrections, particularly as evidenced by abnormal isotope effects on reaction velocities or Arrhenius parameters. A comparison of polar and non-polar solvents would be of particular interest. Hydrogen isotope effects have been reported for a number of proton-transfer reactions in mixtures of water and dipolar aprotic solvents [186–188, 193]. These reveal modest variations in the isotope effects with solvent composition, usually bearing no relation to the solvent effects on the reaction velocity or on the position of equilibrium. The results can be interpreted in terms of changes of solvation during the sequence reactants → transition state → products, but none of the reactions investigated shows unequivocal signs of tunnelling, so that little light is thrown on the present problem. The only clear evidence for solvent effects on tunnelling comes from the work of Caldin and his collaborators [324, 196, 326, 346–349] on the reaction between 4-nitrophenylnitromethane and tetramethylguanidine in a variety of aprotic solvents. The results are given in Tables 5.3 and 5.4 and were discussed briefly in Section 5.5. The behaviour in all the eight solvents used is characteristic of appreciable tunnel corrections, in that the values of $k^H/k^D$, $E_A^D - E_A^H$ and $A_A^D/A_A^H$ all exceed the maximum semi-classical values. However, there are large variations between the different solvents, which fall into two groups. In acetonitrile, dichloromethane and tetrahydrofuran $k^H/k^D$ at 298 K is in the range 11–13, while in the group of less polar solvents (cyclohexene, mesitylene, toluene, dibutyl ether, anisole and chlorobenzene) it is much higher, ranging between 31 and 50. A similar distinction between the two groups of solvents appears in the isotope effects on the Arrhenius parameters (Table 5.4). As shown in Table 5.5 similar differences are naturally found in the barrier dimensions deduced on the basis of a parabolic barrier with the usual assumption that the tunnelling particle has a mass of $m_H$ or $m_D$, independent of the solvent. It seems unreasonable that a change of solvent could produce these relatively large changes in the true width and curvature of the reaction barrier. It was therefore suggested by Caldin and Mateo [196] that in the more polar solvents the effective mass of the proton or deuteron is increased by coupling with the rotation of the polar solvent molecules, thus leading to a smaller tunnel correction. In the less polar media, on the other hand, the proton–solvent interaction involves mainly electron polarization, which does not imply any significant increase in effective mass. The same distinction between dipolar and electronic polarization had previously been pointed out by Kurz [359], who concluded (though without experimental evidence) that tunnelling in proton-transfer reactions should be favoured by solvents of low polarity and dielectric constant.

Caldin and Mateo [196] therefore analysed their results by a different procedure, in which the effective masses $m'_H$ and $m'_D$ were regarded as disposable parameters which were optimized by the computer program.* The resulting effective masses and barrier characteristics are shown in Table 7.1. They may be compared with the corresponding entries in Table 5.5, which are based on the assumption that $m'_H = 1$ a.m.u., $m'_D = 2$ a.m.u., independent of solvent.

The application of the same treatment to the results for the five less polar solvents gave $m'_H = 1.00 \pm 0.01$ a.m.u. and $m'_D = 2.00 \pm 0.01$ a.m.u. throughout, so that for these solvents the barrier dimensions are as given in Table 5.5. It is noteworthy that the barrier dimensions for the two sets of solvents are now very similar, as would be expected on physical grounds; moreover, values derived for $m'_H$ and $m'_D$ are of the same order of magnitude as those deduced from an electrostatic model (Equation 7.5). Similar behaviour has been observed for the reactions of 4-nitrophenylnitromethane with other bases [323, 324]. The suggestion that little displacement of solvent molecules occurs when these reactions take place in solvents of low polarity is supported by the observation [562] that the volume of activation (determined from kinetic measurements at high pressures) has very similar values in five solvents of this type, though corresponding measurements have not been reported for polar solvents.

The observed solvent dependence of tunnel corrections in the reactions of 4-nitrophenylnitromethane with bases can thus be reasonably attributed to

---

* Only one disposable mass parameter was in fact involved, since the analysis assumes that $m'_D - m'_H = 1$ a.m.u. This is not a self-evident assumption, but it is consistent with a simple electrostatic model (Equation 7.5), and it is not likely that any alternative treatment would alter the picture appreciably.

# A review of the present position

Table 7.1. *Effective particle masses and barrier characteristics for the reaction of 4-nitrophenylnitromethane with tetramethylguanidine in polar solvents*

| Solvent | $m'_H$ (a.m.u.) | $m'_D$ (a.m.u.) | $a$ (pm) | $v_{\ddagger}^H$ (cm$^{-1}$) |
|---|---|---|---|---|
| tetrahydrofuran | 1.17 | 2.17 | 78 | 1062 |
| dichloromethane | 1.24 | 2.24 | 79 | 1008 |
| acetonitrile | 1.27 | 2.27 | 79 | 956 |

variations in the effective mass of the proton arising from coupling between proton and solvent motions. However, further experimental evidence is desirable, especially as there have been some differences of opinion [346–349] about the detailed interpretation of the kinetic observations.

A much more radical view of the role of polar solvents in proton-transfer reactions is basic to the theory of Dogonadze and his co-workers, which has also been applied to the cathodic discharge of hydrogen [493–499]. In addition to a series of detailed papers [563–570] a general account of the application of the theory to proton transfers in solution has been published recently [571]. These views have not been generally accepted, and only a brief account will be given here.

Like the Marcus formulation (Equation 7.3) the Dogonadze treatment derives from theories of electron-transfer processes, and assumes a complete separation between the actual proton transfer on the one hand and solvent reorganization (and other movements of heavy atoms) on the other. It goes further by supposing that the actual proton transfer is not an activated process, but takes place by a purely tunnelling mechanism from the lowest possible vibrational level after thermal fluctuations have brought the solvent molecules or the molecular framework into a configuration favourable to tunnelling. The activation energy is thus associated primarily with solvent reorganization rather than with the state of the reactants. The energy profile for the tunnelling process is represented by two intersecting potential energy curves (usually parabolae or Morse curves) rather than by the smooth curves usually assumed in treating these problems, and the transition probability is calculated by second-order perturbation theory. By introducing certain approximations and assumptions about the values assigned to parameters this model is able to reproduce several features of the observed behaviour of proton-transfer reactions, for example the dependence of reaction velocity on $\Delta G_0$ for the reaction (i.e. linear and curved Brönsted plots), deviations from the Arrhenius equation at low temperatures, low reaction rates for pseudo-acids, and the dependence of kinetic isotope effects on transition state symmetry. All these features can of course also be accounted for on the basis of other models.

The theory outlined above can be criticized in a number of respects. As discussed in the earlier part of this section, the separate treatment of solvent re-organization and the transfer process is reasonable for electron transfers, but questionable for the transfer of the proton, which is 1800 times heavier.* The values assigned to the solvent re-organization energy also sometimes appear unreasonably high, for example 200–250 kJ mol$^{-1}$ even for reactions of symmetrical charge type. Further, the assumption of tunnelling from a sharply defined vibrational level ignores the evidence for a broader energy distribution due to superposition of translational or low-frequency motions of the reactants, as discussed in Section 7.2.3. It is also not clear why it is legitimate to represent the process in terms of two intersecting potential energy curves. This is appropriate in cases of small overlap, such as electron-transfer reactions. Proton transfers, on the other hand, involve a transition between two covalently bound states, and one would expect a rounded barrier such as that commonly assumed for hydrogen atom transfers.

However, the main arguments against the theory of Dogonadze and his co-workers come from some of its experimental consequences. The special role attributed to the re-organization of the molecules of polar solvents suggests that the kinetic behaviour of proton-transfer reactions in such media should show special features not encountered in non-polar solvents, in gas reactions, or in hydrogen atom transfers. In fact, as shown particularly in Chapter 5, experimental evidence shows that the effect of tunnelling is very similar in all these types of process, and can be reasonably interpreted on a common basis. Moreover, serious discrepancies with other experimental facts are revealed if we examine the values which the Dogonadze treatment has to assume for the quantity $\Delta R$, the distance through which the proton moves in its tunnelling process: these are all in the range 30–50 pm. In the transfer of a proton from AH to B the distance $\Delta R$ will be equal to $r_{AB} - r_{AH} - r_{BH}$, where $r_{AB}$ is the distance between A and B in the configuration for which tunnelling occurs and $r_{AH}$ and $r_{BH}$ are bond lengths. Since in the systems usually considered these bond lengths are close to 100 pm (108 for C–H, 104 for N–H and 96 for O–H) values of $\Delta R$ between 30 and 50 pm imply A———B distances between 230 and 250 pm. These distances are unreasonably small, since in the absence of hydrogen bonding the closest distance of approach of two carbon, nitrogen or oxygen atoms is in the range 300–350 pm. Even in a very strong hydrogen bond such as that in ice the O———O distance is only reduced to 276 pm, and for the much weaker hydrogen bonds involving C–H the distance between the two centres is much greater. For example, most C–H———O bonds have C———O distances between 310 and 330 pm, with only a few

* Dogonadze et al. make a sharp distinction between quantum and classical degrees of freedom, for which $h\nu/kT > 1$ and $< 1$, respectively. The basis for this distinction is not clear, and in any case the times required for small rotational or translational movements of water molecules [572–574] are little greater than those of proton vibrations, especially if bending modes are included.

examples down to 280 pm [575]. Since the repulsion energy between two non-bonded atoms increases very rapidly as they approach, it appears that the attainment of A——B distances between 230 and 250 pm, as required by the Dogonadze treatment, would demand so much energy that the unfavourable Boltzmann factor would outweigh by far any increase in tunnelling probability.

The crucial importance of the proton-transfer distance $\Delta R$ appears with special clarity when considering kinetic isotope effects, since it is here that low values of $\Delta R$ are particularly required in order to reproduce experimental results. We shall consider as an example proton transfers from a >$CH_2$ group to the group $^-O$- in an oxyanion, since this is a case frequently encountered in practice. On the basis of a simplified model German and Dogonadze [571] give the following equation for the isotope effect at 300 K,

$$\frac{k^H}{k^D} = \exp[1.534 \times 10^{-3}(\Delta R)^2] \quad (7.6)$$

where $\Delta R$ is expressed in pm and the frequency of the C–H vibrations has been taken as 2800 cm$^{-1}$.* The values calculated from this expression are shown in Table 7.2 as a function of $\Delta R$. The table also gives the distance $r$ between the centre of the >$CH_2$ group and the oxygen atom, given by $r$ (pm) = $\Delta R$ + 108 + 96, where the last two figures represent the C–H and O–H bond lengths, respectively.

Since the values observed for $k^H/k^D$ in this class of reaction are normally between 3 and 10 at ordinary temperatures, it is clear that the Dogonadze treatment requires values of $\Delta R$ in the range 30–40 pm, and hence values of $r$ of 234–244 pm. The last column of Table 7.2 contains the estimated interaction energy $V(r)$ (mostly repulsive) between >$CH_2$ and $^-O$- as a function of $r$, taking as a model the interaction between a methane molecule and a neon atom. The calculation is based upon the Lennard–Jones potential (Equation 7.7), where $\sigma$ is the distance at which the attractive and repulsive energies add up to zero, and $\varepsilon$ is the depth of the energy minimum, which occurs at a separation of $2^{1/6} \sigma$.

$$V(r) = 4\varepsilon[(\sigma/r)^{12} - (\sigma/r)^6]. \quad (7.7)$$

For methane–methane and neon–neon interactions the parameters in Equation 7.7 can be determined from an analysis of the second virial

---

* This expression assumes that $\Delta R$ has the same value for the two isotopes. This is not strictly correct, since the variation of tunnelling probability with distance depends on the isotopic mass. However, for a steeply rising repulsive potential this difference is very small. Somewhat different results are obtained if bending vibrations are considered instead of stretching ones, or if the parabolic potentials on which Equation 7.6 is based are replaced by Morse curves, but the general conclusions remain the same.

Table 7.2. *Tunnelling isotope effects and interaction energies for the reaction* $>CH_2 + {}^-O-$

| $r$ (pm) | $\Delta R$ (pm) | $k^H/k^D$ | $V(r)$ (kJ mol$^{-1}$) |
|---|---|---|---|
| 350 | 146 | $1.55 \times 10^{14}$ | $-0.50$ |
| 340 | 136 | $2.06 \times 10^{12}$ | $-0.34$ |
| 330 | 126 | $3.71 \times 10^{10}$ | 0 |
| 320 | 116 | $9.09 \times 10^{8}$ | 0.59 |
| 310 | 106 | $3.02 \times 10^{7}$ | 1.59 |
| 300 | 96 | $1.37 \times 10^{6}$ | 3.39 |
| 290 | 86 | $8.39 \times 10^{4}$ | 6.11 |
| 280 | 76 | $7.01 \times 10^{3}$ | 10.8 |
| 270 | 66 | 795 | 18.7 |
| 260 | 56 | 122 | 32.0 |
| 250 | 46 | 25.6 | 54.7 |
| 240 | 36 | 7.29 | 93.7 |
| 230 | 26 | 2.82 | 162.2 |

coefficients of these gases, which yields the values $\varepsilon/k = 148$ K, $\sigma = 429$ pm for methane, and $\varepsilon/k = 34.9$ K, $\sigma = 278$ pm for neon [576]. The parameters for methane–neon interaction were obtained from the generally accepted combination rules $\sigma_{12} = \frac{1}{2}(\sigma_{11} + \sigma_{22})$, $\varepsilon_{12} = (\varepsilon_{11}\varepsilon_{22})^{1/2}$, which give $\varepsilon/k = 71.9$ K, $\sigma = 330$ pm.

It is apparent that the repulsion energies for values of $\Delta R$ between 30 and 40 pm are so high as to make these configurations highly improbable at readily accessible temperatures. In fact, a more detailed treatment [577] shows that for the model used here the reaction probability at 300 K reaches a maximum at $r = 280$ pm, $\Delta R = 76$ pm, falling off rapidly at both larger and smaller separations. For this configuration Equation 7.6 predicts $k^H/k^D = 7000$, compared with the commonly observed values in the range 3–10. These results would be modified quantitatively by making different assumptions about the dependence on separation of the tunnelling probability and the interaction energy, but the conclusion seems inescapable that the Dogonadze theory cannot be made to agree with experiment on any realistic assessment of the repulsive forces. Even if a closer approach between reactants were possible in some instances there would certainly be some systems in which $\Delta R$ would exceed 60 pm because of repulsion between other parts of the reactants (steric hindrance). This should lead to isotope effects of at least several hundred: as shown in Table 5.3 the highest value reported for a proton-transfer reaction in a polar solvent is 27. On the other hand the value $\Delta R = 76$ pm quoted above as being the most probable distance for proton transfer is close to the barrier widths derived from experimental results by the more conventional interpretation of tunnel corrections (cf. Tables 5.5 and 7.1).

There are thus serious objections to the theories of Dogonadze and his

# A review of the present position

collaborators in their present form. It is probable that the sharp separation between solvent re-organization and proton tunnelling needs modifying by introducing some coupling between these processes and by allowing a spread of energies for the proton, thus bringing the approach more into line with the treatment adopted in this book. On the other hand, there is no doubt that most current analyses of tunnelling in proton-transfer reactions in polar media suffer a serious weakness in their neglect of the part played by solvent re-organization. Fortunately this problem does not arise for reactions in non-polar media or in the gas phase, or for processes involving the transfer of hydrogen atoms rather than protons, all of which provide unequivocal evidence of the frequent importance of the tunnel effect.

# Appendix A
# Notes on hypergeometric functions

Hypergeometric functions have been studied extensively for more than two hundred years: they form the subject of several books and of at least one chapter in most books on analysis or differential equations. We shall give here only a few definitions and properties which are used in other parts of this book.

The original definition of the hypergeometric function was in terms of the series

$$F(\alpha, \beta, \gamma, z) = 1 + \frac{\alpha \beta}{\gamma} \frac{z}{1!} + \frac{\alpha(\alpha+1)\beta(\beta+1)z^2}{\gamma(\gamma+1)\ 2!} + \cdots \qquad (A.1)$$

where $\alpha$ and $\beta$ can have any values, but the values $\gamma = 0, -1, -2$, etc., are excluded. This series is convergent for real values of $z$ less than unity and also for complex values when $|z| < 1$. Direct substitution shows that $F(\alpha, \beta, \gamma, z)$ is a particular solution of the hypergeometric equation

$$z(1-z)\frac{d^2u}{dz^2} + [\gamma - (\alpha + \beta + 1)z]\frac{du}{dz} - \alpha\beta u = 0. \qquad (A.2)$$

Since Equation A.2 is a second-order differential equation there must exist a second linearly independent solution, which is found to be

$$z^{1-\gamma} F(\beta - \gamma + 1, \alpha - \gamma + 1, 2 - \gamma, z)$$

so that the general solution of Equation A.2 is

$$u = A_1 F(\alpha, \beta, \gamma, z) + A_2 z^{1-\gamma} F(\beta - \gamma + 1, \alpha - \gamma + 1, 2 - \gamma, z) \qquad (A.3)$$

where $A_1$ and $A_2$ are arbitrary constants. This result is used in Section 2.4.1 for deriving the permeability of the Eckart barrier.

The series in Equation A.1 is not applicable when $|z| > 1$, since it diverges, but convergent series for this range of the variable can be obtained from various transformation equations, of which the most useful for present purposes is

$$F(\alpha, \beta, \gamma, z) = \frac{\Gamma(\gamma)\Gamma(\beta - \alpha)}{\Gamma(\beta)\Gamma(\gamma - \alpha)}(-z)^{-\alpha} F\left(\alpha, \alpha + 1 - \gamma, \alpha + 1 - \beta, \frac{1}{z}\right)$$

$$+ \frac{\Gamma(\gamma)\Gamma(\alpha - \beta)}{\Gamma(\alpha)\Gamma(\gamma - \beta)}(-z)^{-\beta} F\left(\beta, \beta + 1 - \gamma, \beta + 1 - \alpha, \frac{1}{z}\right) \qquad (A.4)$$

# Appendix A

Since now $|1/z| < 1$, Equation A.1 can be used for expanding the two hypergeometric functions on the right-hand side of Equation A.4. This procedure is used in Appendix C for deriving an expression for the tunnel correction to the reaction velocity for a parabolic barrier, and is also referred to in Section 2.4.1 in connection with the asymptotic behaviour of the wave functions for the Eckart barrier.

In the general case where $\alpha$ or $\beta$ is non-integral Equation A.4 contains an ambiguity because of the many-valued nature of $(-z)^{-\alpha}$ or $(-z)^{-\beta}$: for example, $1^{1/3}$ can have the values $1$, $-\frac{1}{2}(1-\sqrt{3}i)$ or $-\frac{1}{2}(1+\sqrt{3}i)$. The rule here is to take the value having the lowest absolute value of the argument, i.e. of the angle $\theta$ when the complex quantity is written in the form $r(\cos\theta + i\sin\theta)$. The argument itself is not unique, since the value of the last expression is unchanged if $\theta$ is replaced by $\theta + 2n\pi$, where $n$ is an integer: we are concerned here with the principal value, which is defined as the value between $-\pi$ and $+\pi$.

The hypergeometric function can also be expressed in terms of an integral. In the general case this is a contour integral in the complex plane. When only real quantities are concerned the integral becomes

$$F(\alpha, \beta, \gamma, z) = \frac{\Gamma(\gamma)}{\Gamma(\beta)\Gamma(\gamma - \beta)} \int_0^1 t^{\beta-1}(1-t)^{\gamma-\beta-1}(1-tz)^{-\alpha} dt \quad (A.5)$$

which is valid provided that $\gamma > \beta > 0$. Direct substitution shows that the integral satisfies the hypergeometric Equation A.2, while the multiplying constant is chosen so as to give the function a value of unity when $z = 0$. A special case of Equation A.5 is used in Appendix C in deriving the tunnel correction to the reaction velocity for a parabolic barrier.

The *confluent hypergeometric function* $F(\alpha, \gamma, z)$ is defined by the series

$$F(\alpha, \gamma, z) = 1 + \frac{\alpha}{\gamma}\frac{z}{1!} + \frac{\alpha(\alpha+1)}{\gamma(\gamma+1)}\frac{z^2}{2!} + \cdots \quad (A.6)$$

which is convergent for all finite values of $|z|$ and satisfies the differential equation

$$z^2\frac{d^2u}{dz^2} + (\gamma - z)\frac{du}{dz} - \alpha u = 0. \quad (A.7)$$

Comparison of Equations A.1 and A.6 shows that it is obtained from the hypergeometric function by taking the limit

$$F(\alpha, \gamma, z) = \lim_{\beta \to \infty} F(\alpha, \beta, \gamma, z/\beta). \quad (A.8)$$

Substitution shows that a second, linearly independent solution of Equation A.7 is

$$u = z^{1-\gamma} F(\alpha - \gamma + 1, 2 - \gamma, z)$$

so that the general solution is

$$u = A_1 F(\alpha, \gamma, z) + A_2 z^{1-\gamma} F(\alpha - \gamma + 1, 2 - \gamma, z) \quad (A.9)$$

where $A_1$ and $A_2$ are constants.

For large values of $|z|$ the series in Equation A.6 converges slowly, and use can be made of the asymptotic expansion

$$F(\alpha, \gamma, z) = \frac{\Gamma(\gamma)}{\Gamma(\gamma - \alpha)}(-z)^{-\alpha} G(\alpha, \alpha - \gamma + 1, -z)$$

$$+ \frac{\Gamma(\gamma)}{\Gamma(\alpha)} e^z z^{\alpha - \gamma} G(\gamma - \alpha, 1 - \alpha, z) \quad (A.10)$$

in which

$$G(\alpha, \beta, z) = 1 + \frac{\alpha \beta}{1!z} + \frac{\alpha(\alpha + 1)\beta(\beta + 1)}{2!z^2} + \cdots \quad (A.11)$$

As in the asymptotic expansion of the hypergeometric function (Equation A.4) if $\alpha$ or $\alpha - \gamma$ are non-integral $(-z)^{-\alpha}$ and $z^{\alpha - \gamma}$ are many-valued: the value having the lowest absolute value of the argument is again to be taken.

The confluent hypergeometric function can also be represented by integrals in various ways, for example

$$F(\alpha, \gamma, z) = \frac{\Gamma(\gamma)}{\Gamma(\alpha)\Gamma(\gamma - \alpha)} z^{1-\gamma} \int_0^z e^t t^{\alpha - 1}(z - t)^{\gamma - \alpha - 1} dt. \quad (A.12)$$

The confluent hypergeometric function has many physical applications. In the present book it occurs in calculating barrier permeabilities for the double anharmonic potential (Section 2.4.3) and for the inverted Morse potential (Section 2.4.4).

# Appendix B
# Permeabilities for the barrier $V(x) = V_0\{1 - |(x/b)^n|\}$

As stated in Section 2.4.5, explicit expressions for the permeability of the above barrier when $n$ has an arbitrary positive value can only be obtained when $W = V_0$. It is convenient to write $n = 2q - 2$ $(1 < q < \infty)$, when the Schrödinger equation becomes

$$\frac{d^2\psi}{dx^2} + \frac{2mV_0}{\hbar^2}\left|\left(\frac{x}{b}\right)^{2q-2}\right|\psi = 0. \tag{B.1}$$

The substitutions

$$\xi = \beta^{1/q}x/b, \quad \beta = (2mV_0)^{1/2}b/\hbar \tag{B.2}$$

convert Equation B.1 into

$$\frac{d^2\psi}{d\xi^2} + |\xi|^{2q-2}\psi = 0 \tag{B.3}$$

which is known to have the solution*

$$\psi = \xi^{1/2} Z_{\pm 1/2q}\left(\frac{\xi^q}{q}\right) \tag{B.4}$$

where $Z$ represents any type of Bessel function of order $\pm 1/2q$. The general solution consists of a linear combination of any two such independent functions, and the permeability of an infinite barrier of the type considered can be derived by choosing these functions and their coefficients so that as $\xi \to \infty$ the wave function represents only an outgoing wave, as was done in treating the Eckart potential (Section 2.4.1). Alternatively, the barrier may be truncated on either side by regions of constant potential energy, and a general expression for the permeability obtained by applying the equations of continuity of $\psi$ and $d\psi/dx$ between the three regions. Since we are considering only particles of energy $V_0$ the expression for an infinite barrier can be derived

---

* This result, and the relations involving Bessel functions and gamma functions used in this section, may be found in any standard work on analysis.

by going to the limit of an infinitely wide and high barrier. The second of these approaches will be adopted here.

We suppose that the barrier is truncated at $x = \pm b$ ($\xi = \pm \beta^{1/q}$) by regions of constant (zero) potential energy.* The wave functions in the three regions can then be written as follows:

$$\left.\begin{aligned} &-\infty < \xi < -\beta^{1/q} \\ &\psi_{\mathrm{I}} = \exp[i(2mV_0)^{1/2}x/\hbar] + C\exp[-i(2mV_0)^{1/2}x/\hbar] \\ &\quad\;\; = \exp(i\beta^{1-1/q}\xi) + C\exp(-i\beta^{1-1/q}\xi) \\ &-\beta^{1/q} < \xi < +\beta^{1/q} \\ &\psi_{\mathrm{II}} = \xi^{1/2}\left[AJ_{1/2q}\left(\frac{\xi^q}{q}\right) + BJ_{-1/2q}\left(\frac{\xi^q}{q}\right)\right] \\ &+\beta^{1/q} < \xi < \infty \\ &\psi_{\mathrm{III}} = D\exp(i\beta^{1-1/q}\xi) \end{aligned}\right\} \quad (\text{B.5})$$

The coefficients $A$, $B$, $C$ and $D$ are to be determined, and the permeability is then given by $|D|^2$. In applying the conditions of continuity at $\xi = \pm\beta^{1/q}$ we note that use of the general relation

$$\frac{dJ_n(z)}{dz} = \frac{n}{z}J_n(z) - J_{n+1}(z) = -\frac{n}{z}J_n(z) + J_{n-1}(z) \quad (\text{B.6})$$

gives

$$\frac{d\psi_{\mathrm{II}}}{d\xi} = \xi^{q-(1/2)}\left[AJ_{1/2q-1}\left(\frac{\xi^q}{q}\right) - BJ_{1-1/2q}\left(\frac{\xi^q}{q}\right)\right] \quad (\text{B.7})$$

The equations of continuity then become

$$\left.\begin{aligned} e^{-i\beta} + Ce^{i\beta} &= \beta^{1/2q}\left[-AJ_{1/2q}\left(\frac{\beta}{q}\right) + BJ_{-1/2q}\left(\frac{\beta}{q}\right)\right] \\ i\beta^{1-1/q}e^{i\beta} - Ci\beta^{1-1/q}e^{-i\beta} &= \beta^{1-1/2q}\left[AJ_{1/2q-1}\left(\frac{\beta}{q}\right) + BJ_{1-1/2q}\left(\frac{\beta}{q}\right)\right] \\ De^{i\beta} &= \beta^{1/2q}\left[AJ_{1/2q}\left(\frac{\beta}{q}\right) + BJ_{-1/2q}\left(\frac{\beta}{q}\right)\right] \\ Di\beta^{1-1/q}e^{i\beta} &= \beta^{1-1/2q}\left[AJ_{1/2q-1}\left(\frac{\beta}{q}\right) - BJ_{1-1/2q}\left(\frac{\beta}{q}\right)\right] \end{aligned}\right\} \quad (\text{B.8})$$

---

* This might appear to be an arbitrary procedure, but truncation at any other energy can be reduced to the present problem by changing the zero of energy, which does not of course affect the physical situation.

## Appendix B

Eliminating the coefficients $A$, $B$ and $C$ and introducing the relation

$$J_n(z)J_{-n+1}(z) + J_{-n}(z)J_{n-1}(z) = \frac{2\sin n\pi}{\pi z} \tag{B.9}$$

gives finally for the permeability

$$G_0 = |D|^2$$
$$= \frac{4q^2 \sin^2(\pi/2q)}{\pi^2 \beta^2 \left[ J^2_{1/2q}\left(\frac{\beta}{q}\right) + J^2_{1/2q-1}\left(\frac{\beta}{q}\right) \right]\left[ J^2_{-1/2q}\left(\frac{\beta}{q}\right) + J^2_{1-1/2q}\left(\frac{\beta}{q}\right) \right]} \tag{B.10}$$

The corresponding equation for an infinite barrier is obtained by letting both $V_0$ and $b$ tend to infinity, keeping the ratio $V_0/b^{2q-2}$ constant so as to preserve the shape of the barrier. This implies that $\beta \to \infty$ (Equation B.2) and we can therefore use the asymptotic expansions for the Bessel functions,

$$J_n(z) = \frac{2}{\pi z}\{\cos[z - \tfrac{1}{4}(2n+1)\pi]$$

$$- \frac{4n^2 - 1}{8z}\sin[z - \tfrac{1}{4}(2n+1)\pi] + \ldots\} \tag{B.11}$$

whence

$$\lim_{\beta \to \infty}\left[ J^2_{1/2q}\left(\frac{\beta}{q}\right) + J^2_{1/2q-1}\left(\frac{\beta}{q}\right) \right] = \lim_{\beta \to \infty}\left[ J^2_{-1/2q}\left(\frac{\beta}{q}\right) + J^2_{1-1/2q}\left(\frac{\beta}{q}\right) \right]$$

$$= \frac{2q}{\pi\beta} \tag{B.12}$$

and therefore from Equation B.10

$$\lim_{\beta \to \infty} G_0 = \sin^2\frac{\pi}{2q} = \sin^2\frac{\pi}{n+2}. \tag{B.13}$$

This limit applies not only to infinite barriers, but also to truncated barriers of the form considered having sufficiently large dimensions ($V_0$ and $b$), and to the passage of sufficiently heavy particles across any barriers of this form.

Equation B.13 is in fact a good approximation for hydrogen atoms or protons crossing truncated barriers of the dimensions commonly met with in chemical problems. If we include the second term in the asymptotic expansion of Equation B.11, Equation B.10 becomes

$$G_0 = \frac{\sin^2\dfrac{\pi}{2q}}{1 - \dfrac{q-1}{2\beta}\cos\dfrac{\pi}{2q}\cos\dfrac{2\beta}{q}} \tag{B.14}$$

correct to terms in $1/\beta$. For a parabolic barrier ($q = 2$) and values of $\beta$ between 10 and 20 the denominator of Equation B.14 oscillates about unity with maximum and minimum values of 1.035 and 0.965. A proton crossing a barrier of height $7.0 \times 10^{-20}$ J (42 kJ mol$^{-1}$) and width $2 \times 10^{-10}$ m corresponds to $\beta = 14.4$.

If $q \to \infty$ the barrier considered becomes a rectangular barrier of width $2b$. The Bessel functions in Equation B.10 for the permeability are then of order zero or $\pm 1$, and the whole expression becomes indeterminate. The wave function, Equation B.4, is in fact no longer an adequate solution, since $J_{1/2q}$ and $J_{-1/2q}$ cease to be linearly independent. However, the correct value for the permeability may be obtained by writing $1/q = \varepsilon$ and evaluating the limiting value of Equation B.10 for $\varepsilon \to 0$ by means of the relations

$$J_n(z) = \frac{z^n}{2^n \Gamma(z+1)} \left[ 1 - \frac{z^2}{2(2n+2)} + \ldots \right] \tag{B.15}$$

and

$$\Gamma(z)\Gamma(1-z) = \pi/\sin \pi z. \tag{B.16}$$

The result is

$$\lim_{q \to \infty} G_0 = \frac{1}{1+\beta^2} \tag{B.17}$$

which agrees with Equation 2.23 derived previously for the permeability of a rectangular barrier of width $2b$.

# Appendix C
# Derivation of the tunnel correction for a parabolic barrier

The expression to be evaluated (cf. Equation 3.4) is

$$Q_t = \frac{e^\alpha}{kT} \int_0^\infty \frac{\exp(-W/kT)dW}{1+\exp[\beta(E-W)/E]} \quad (C.1)$$

where $\alpha = E/kT$, $\beta = E/\hbar v_\ddagger$ and $iv_\ddagger$ is the imaginary frequency of the particle in the barrier. On making the substitution $x = \exp[\alpha(E-W)/E]$ this becomes

$$Q_t = \int_0^{e^\alpha} \frac{dx}{1+x^{\beta/\alpha}} = \int_0^\infty \frac{dx}{1+x^{\beta/\alpha}} - \int_{e^\alpha}^\infty \frac{x^{-\beta/\alpha}dx}{1+x^{-\beta/\alpha}}. \quad (C.2)$$

The first integral on the right is equal to $(\pi\alpha/\beta)\operatorname{cosec}(\pi\alpha/\beta)$, provided that $\alpha < \beta$. In the last integral the limits of $x^{-\beta/\alpha}$ are 0 and $e^{-\beta}$, so that the denominator can be expanded by the binomial theorem and integration carried out term by term. The result is

$$Q_t = \frac{\pi\alpha/\beta}{\sin(\pi\alpha/\beta)} - \alpha e^\alpha \left( \frac{e^{-\beta}}{\beta-\alpha} - \frac{e^{-2\beta}}{2\beta-\alpha} + \frac{e^{-3\beta}}{3\beta-\alpha} - \ldots \right) \quad (C.3)$$

as given in Equation 3.19. This expression was first derived by Bell in 1959, using the above method [81]. The derivation appears to be valid only for $\alpha < \beta$ (i.e. $hv_\ddagger/kT < 2\pi$), but this restriction can be removed by the following argument.

Any value of $\alpha > \beta$ can be expressed as $\alpha = \alpha_0 + m\beta$, where $\alpha_0 < \beta$, and $m$ is a positive integer. Introducing the variable $y = \exp[\alpha_0(E-W)/E] = x^{\alpha_0/\alpha}$ Equation C.1 or Equation C.2 becomes

$$Q_t(\alpha, \beta) = \frac{\alpha}{\alpha_0} \int_0^{e^{\alpha_0}} \frac{y^{m\beta/\alpha_0}dy}{1+y^{\beta/\alpha_0}}. \quad (C.4)$$

Since by long division

$$\frac{z^m}{1+z} = z^{m-1} - z^{m-2} + z^{m-3} - \ldots + (-1)^{m-1} + \frac{(-1)^m}{1+z}$$

Equation C.4 becomes

$$Q_t(\alpha, \beta) = \frac{\alpha}{\alpha_0}\left\{(-1)^m \int_0^{e^{\alpha_0}} \frac{dy}{1+y^{\beta/\alpha}}\right.$$
$$+ \int_0^{e^{\alpha_0}} [y^{(m-1)\beta/\alpha_0} - y^{(m-2)\beta/\alpha_0} + \ldots + (-1)^{m-1}]dy\}. \tag{C.5}$$

The first integral in Equation C.5 is $Q_t(\alpha_0, \beta)$, and since by definition $\alpha_0 < \beta$ it can be replaced by the right-hand side of Equation C.3, giving

$$Q_t(\alpha, \beta) = (-1)^m \frac{\pi\alpha_0/\beta}{\sin(\pi\alpha_0/\beta)} - (-1)^m \alpha e^{\alpha_0}\left(\frac{e^{-\beta}}{\beta - \alpha_0} - \frac{e^{-2\beta}}{2\beta - \alpha_0} + \ldots\right)$$
$$- \alpha e^{\alpha_0}\left[\frac{e^{(m-1)\beta}}{\beta - \alpha} - \frac{e^{(m-2)\beta}}{2\beta - \alpha} + \ldots + \frac{(-1)^m}{\alpha_0}\right]. \tag{C.6}$$

Since $\alpha = \alpha_0 + m\beta$, $(-1)^m \sin(\pi\alpha_0/\beta) = \sin(\pi\alpha/\beta)$, and the first series in Equation C.6 is a continuation of the second. The whole expression thus becomes identical with Equation C.3, which is therefore valid for any values of $\alpha$ and $\beta$.

A neater derivation of Equation C.3, valid for all values of $\alpha$ and $\beta$, makes use of the properties of the hypergeometric function, summarized in Appendix A. If we write $v = \exp[\beta(E - W)/E] = x^{\beta/\alpha}$, $\mu = \alpha/\beta$, then Equation C.1 or Equation C.2 becomes

$$Q_t = \mu \int_0^{e^\beta} \frac{v^{\mu-1} dv}{1+v}. \tag{C.7}$$

Since $\Gamma(1) = 1$, $\Gamma(x+1) = x\Gamma(x)$, a special case of the integral representation of the hypergeometric function (Equation A.5) is

$$F(1, \mu, \mu+1, -z) = \mu \int_0^1 \frac{t^{\mu-1} dt}{1+tz} = \frac{\mu}{z^\mu} \int_0^z \frac{v^{\mu-1} dv}{1+v}. \tag{C.8}$$

Comparison of Equations C.7 and C.8 gives

$$Q_t = e^\alpha F(1, \mu, \mu+1, -e^\beta). \tag{C.9}$$

Since $|-e^\beta| > 1$, Equation C.9 cannot be evaluated by using the usual expansion of the hypergeometric function (Equation A.1), but it can be transformed into a function of $-e^{-\beta}$ by using Equation A.4. The result is

$$Q_t = \frac{\Gamma(\mu+1)\Gamma(1-\mu)}{[\Gamma(1)]^2} F(\mu, 0, \mu, -e^{-\beta})$$
$$+ \frac{\Gamma(\mu+1)\Gamma(\mu-1)}{[\Gamma(\mu)]^2} e^{\alpha-\beta} F(1, 1-\mu, 2-\mu, -e^{-\beta}). \tag{C.10}$$

## Appendix C

This can be simplified by using the relations $\Gamma(1) = 1$, $\Gamma(x+1) = x\Gamma(x)$, $\Gamma(x)\Gamma(1-x) = \pi/\sin \pi x$, and the fact that $F(a, b, c, z) = 1$ if either $a$ or $b$ is zero, leading to

$$Q_t = \frac{\pi\mu}{\sin \pi\mu} - \frac{\mu e^{\alpha - \beta}}{1 - \mu} F(1, 1 - \mu, 2 - \mu, -e^{-\beta}). \tag{C.11}$$

The hypergeometric function in Equation C.11 can now be replaced by the usual series (in Equation A.1), which gives

$$F(1, 1 - \mu, 2 - \mu, -e^{-\beta}) = 1 - \frac{1-\mu}{2-\mu}e^{-\beta} + \frac{1-\mu}{3-\mu}e^{-2\beta} - \ldots \tag{C.12}$$

and since $\mu = \alpha/\beta$ we again obtain Equation C.3.

If $\alpha/\beta = m$, where $m$ is an integer, Equation C.3 appears to become indeterminate, since both the term in $\sin(\pi\alpha/\beta)$ and the $m$th term of the series become infinite. However, it is easy to show that the sum of these terms remains finite. If $\alpha/\beta = m(1 + \varepsilon)$, the two offending terms become

$$(-1)^m \frac{\pi m(1+\varepsilon)}{\sin(m\pi\varepsilon)} + (-1)^{m+1}\frac{(1+\varepsilon)e^{m\beta\varepsilon}}{\varepsilon}.$$

Expanding the sine and the exponential as far as the terms in $\varepsilon$ gives

$$(-1)^m\left(\frac{1}{\varepsilon} + 1\right) + (-1)^{m+1}\left(\frac{1}{\varepsilon} + 1 + \alpha\right) = (-1)^{m+1}\alpha. \tag{C.13}$$

These two terms should therefore be replaced by $(-1)^{m+1}\alpha$ when $\alpha$ is close to $m\beta$.

There are a few special cases in which simple exact expressions for $Q_t$ can be obtained, either by using Equations C.3 and C.13, or by direct elementary integration of Equation C.2. Three such results are

$$\left.\begin{array}{ll} \alpha = \beta, & Q_t = \ln(1 + e^{\alpha}) \\ \alpha = 2\beta, & Q_t = 2e^{\alpha/2} - \alpha - 2\ln(1 + e^{-\alpha/2}) \\ \alpha = \tfrac{1}{2}\beta, & Q_t = \tan^{-1} e^{\alpha}. \end{array}\right\} \tag{C.14}$$

# References

[1] De Broglie, L. (1925), *Ann. Phys.*, **10**, 22.
[2] Heisenberg, W. K. (1927), *Z. Phys.*, **43**, 172.
[3] See e.g. Born, M. and Wolf, E. (1965), *Principles of Optics* (Pergamon, Oxford) p. 57.
[4] Newton, Sir Isaac (1704), *Opticks* (pp. 194, 205 in 1952 reprint, Dover, New York).
[5] Coon, D. D. (1966), *Amer. J. Phys.*, **34**, 240.
[6] McDonald, W. J., Udey, S. N., and Hickson, P. (1971), *Amer. J. Phys.*, **39**, 1141.
[7] Castro, J. C. (1975), *Amer. J. Phys.*, **43**, 107.
[8] Feynman, R. P., Leighton, R. B., and Sands, M. (1964), *The Feynman Lectures on Physics*, Vol. II, Sections 33-36 (Addison-Wesley, Reading, Mass.).
[9] Pohl, H. A. (1967), *Quantum Mechanics for Science and Engineering* (Prentice-Hall, New Jersey). p. 50.
[10] Goos, F. and Hänchen, M. (1947), *Ann. de Phys.*, **1**, 333.
[11] Chin, K. W. and Quinn, J. J. (1972), *Amer. J. Phys.*, **40**, 1847.
[12] Lotsch, H. K. V. (1970-71), *Optik (Stuttgart)*, **32**, 116, 189, 299, 553.
[13] Hirschfelder, J. O., Christoph, A. C., and Palke, W. E. (1974), *J. Chem. Phys.*, **61**, 5435.
[14] Oppenheimer, J. R. (1928), *Phys. Rev.*, **31**, 66.
[15] Fowler, R. H. and Nordheim, L. (1928), *Proc. Roy. Soc. A*, **119**, 173.
[16] Nordheim, L. (1928), *Proc. Roy. Soc. A*, **121**, 626.
[17] Duke, C. B. (1969), *Tunneling in Solids* (Academic Press, London and New York).
[18] Burstein, E. and Lundqvist, S. (ed.) (1969), *Tunneling Phenomena in Solids* (Plenum Press, New York).
[19] Solymar, L. (1972), *Superconductive Tunnelling and Applications* (Chapman and Hall, London).
[20] Ziman, J. M. (1972), *Principles of the Theory of Solids*, 2nd edn., Chapters 6 and 11 (Cambridge University Press).
[21] Jaklevic, R. C. and Lambe, J. (1966), *Phys. Rev. Lett.*, **17**, 1139.
[22] Lambe, J. and Jaklevic, R. C. (1968), *Phys. Rev.*, **165**, 821.
[23] Simonsen, M. G. and Coleman, R. V. (1973), *Phys. Rev.*, B, **8**, 5875.
[24] Simonsen, M. G., Coleman, R. V., and Hansma, P. K. (1974), *J. Chem. Phys.*, **61**, 3789.

# References

[25] Brown, N. M. D. and Walmsley, D. G. (1976), *Chem. in Britain*, **12**, 92.
[26] Gamow, G. (1928), *Z. Phys.*, **51**, 204.
[27] Gurney, R. W. and Condon, E. U. (1928), *Nature*, **122**, 439.
[28] Gurney, R. W. and Condon, E. U. (1929), *Phys. Rev.*, **33**, 127.
[29] Geiger, H. and Nuttall, J. M. (1911), *Phil. Mag.*, **22**, 613; (1912), *Phil. Mag.*, **23**, 429.
[30] Friedlander, G., Kennedy, J. W., and Miller, J. M. (1964), *Nuclear and Radiochemistry* (Wiley, New York) pp. 222, 227.
[31] Hund, F. (1927), *Z. Phys.*, **43**, 805.
[32] Oppenheimer, J. R. (1928), *Phys. Rev.*, **31**, 66.
[33] Bourgin, D. G. (1929), *Proc. Nat. Acad. Sci.*, **15**, 357.
[34] Langer, R. M. (1929), *Phys. Rev.*, **34**, 92.
[35] Born, M. and Franck, J. (1930), *Nachr. Ges. Wiss. Göttingen*, 77.
[36] Roginsky, S. and Rosenkewitsch, L. (1930), *Z. Phys. Chem. B*, **10**, 47; (1931), *Z. Phys. Chem. B*, **15**, 103.
[37] Born, M. and Weisskopf, V., (1931), *Z. Phys. Chem. B*, **12**, 206.
[38] Wigner, E. (1932), *Z. Phys. Chem. B*, **19**, 203.
[39] Cremer, E. and Polanyi, M. (1932), *Z. Phys. Chem. B*, **19**, 443.
[40] Bell, R. P. (1933), *Proc. Roy. Soc. A*, **139**, 466.
[41] Bawn, C. E. H. and Ogden, G. (1934), *Trans. Faraday Soc.*, **30**, 432.
[42] Landau, L. D. and Lifshitz, E. M. (1965), *Quantum Mechanics, Non-relativistic Theory*, 2nd edn. (Pergamon Press, London).
[43] Johnston, H. S. (1966), *Gas Phase Reaction Rate Theory* (Ronald Press, New York) p. 337.
[44] Hill, D. L. and Wheeler, J. A. (1953), *Phys. Rev.*, **89**, 1102.
[45] Eckart, C. (1930), *Phys. Rev.*, **35**, 1303.
[46] Duke, C. B. (1969), *Tunneling in Solids* (Academic Press, London and New York) p. 31.
[47] Ter Haar, D. (1964), *Selected Problems in Quantum Mechanics* (Academic Press, New York) p. 129.
[48] Goldman, I. I., Krivchenkov, U. D., Kogan, V. I., Galitzkii, V. M. (1960), *Problems in Quantum Mechanics* (Academic Press, New York) p. 63.
[49] Quickert, K. A. and Le Roy, D. J. (1970), *J. Chem. Phys.*, **52**, 856.
[50] Bell, R. P. (1978), *J. Chem. Soc. Faraday Trans. II*, **74**, 688.
[51] Bell, R. P. (1937), *Proc. Roy. Soc., A*, **150**, 128.
[52] Le Roy, R. J., Quickert, K. A., and Le Roy, D. J. (1970), *Trans. Faraday Soc.*, **66**, 2997.
[53] Le Roy, R. J., Quickert, K. A., and Le Roy, D. J., *University of Wisconsin, Theoretical Chemistry Institute Report* WIS-TCI-384 (1970).
[54] Le Roy, R. J., Sprague, E. D., and Williams, F. (1972), *J. Phys. Chem.*, **76**, 546.
[55] Brillouin, L. (1926), *C.R. Acad. Sci.*, **153**, 24.
[56] Brillouin, L. (1926), *J. de Physique*, **7**, 353.
[57] Wentzel, G. (1926), *Z. Phys.*, **38**, 518.
[58] Kramers, H. A. (1926), *Z. Phys.*, **39**, 828.
[59] Jeffreys, H. (1923), *Proc. London Math. Soc.*, (2), **23**, 428.
[60] Kemble, E. C. (1935), *Phys. Rev.*, **48**, 549.

[61] Kemble, E. C. (1937), *Fundamental Principles of Quantum Mechanics*, (McGraw-Hill, New York) §21 and Appendix D.
[62] Titchmarsh, E. C. (1946), *Eigenfunction Expansions Associated with Second-Order Differential Equations* (Oxford University Press) Chapter VIII.
[63] Fröman, N. and Fröman, P. O. (1965), *The JWKB Approximation* (North Holland Publishing Co., Amsterdam).
[64] Dunham, J. L. (1932), *Phys. Rev.*, **41**, 713, 721.
[65] Jeffreys, B. (1942), *Proc. Camb. Phil. Soc.*, **38**, 401.
[66] Christov, S. G. (1963), *Ann. Phys. (Leipzig)*, **12**, 20.
[67] Berry, R. S. (1960), *Rev. Mod. Phys.*, **32**, 447.
[68] Bell, R. P. (1935), *Phil. Mag.* [7], **25**, 488.
[69] Von Laue, M. (1928), *Z. Phys.*, **52**, 726.
[70] Kemble, E. C. (1937), *Fundamental Principles of Quantum Mechanics* (McGraw-Hill, New York) §31.
[71] Dennison, D. M. and Uhlenbeck, G. E. (1932), *Phys. Rev.*, **41**, 313.
[72] Manning, M. F. (1935), *J. Chem. Phys.*, **3**, 136.
[73] Harmony, M. D. (1971), *Chem. Phys. Lett.*, **10**, 337.
[74] Löwdin, P. O. (1964), *Biopolymers Symp.*, **1**, 161.
[75] Brickmann, J. and Zimmermann, H. (1969), *J. Chem. Phys.*, **50**, 1608.
[76] Brickmann, J. and Zimmermann, H. (1966), *Ber. Bunsenges. Phys. Chem.*, **70**, 157.
[77] Somorjai, R. L. and Hornig, D. F. (1962), *J. Chem. Phys.*, **36**, 1980.
[78] Flanigan, M. C. and de la Vega, J. R. (1974), *J. Chem. Phys.*, **61**, 1882.
[79] Busch, J. H. and de la Vega, J. R. (1977), *J. Amer. Chem Soc.*, **99**, 2397.
[80] Harmony, M. D. (1972), *Chem. Soc. Reviews*, **1**, 211.
[81] Bell, R. P. (1959), *Trans. Faraday Soc.*, **55**, 1.
[82] Johnston, H. S. (1966), *Gas Phase Reaction Rate Theory* (Ronald Press, New York) p. 135.
[83] Johnston, H. S. (1966), *Gas Phase Reaction Rate Theory* (Ronald Press, New York) Appendix C.
[84] Johnston, H. S. (1966), *Gas Phase Reaction Rate Theory* (Ronald Press, New York) p. 51.
[85] Johnston, H. S. (1966), *Gas Phase Reaction Rate Theory* (Ronald Press, New York) p. 195.
[86] Truhlar, D. G. and Kuppermann, A. (1971), *J. Amer. Chem. Soc.*, **93**, 1840.
[87] Karplus, M., Porter, R. N., and Sharma, R. D. (1965), *J. Chem. Phys.* **34**, 3259.
[88] Laidler, K. J., (1969), *Theories of Chemical Reaction Rates* (McGraw-Hill, New York) p. 165.
[89] Johnston, H. S. and Rapp, D. (1961), *J. Amer. Chem. Soc.*, **83**, 1.
[90] Chapman, S., Garrett, B. C., and Miller, W. H. (1975), *J. Chem. Phys.*, **63**, 2710.
[91] Marcus, R. A. (1967), *Disc. Faraday Soc.*, **44**, 167.
[92] Dogonadze, R. R. and Kuznetsov, A. M. (1975), *Progr. Surface Sci.*, **6**, 3.
[93] Bell, R. P. (1935), *Proc. Roy. Soc.*, A, **148**, 241.
[94] Johnston, H. S. (1966), *Gas Phase Reaction Rate Theory* (Ronald Press, New York) p. 44.
[95] Johnston, H. S. and Heicklen, J. (1962), *J. Phys. Chem.*, **66**, 532.

## References

[96] Shin, H. (1963), *J. Chem. Phys.*, **39**, 2934.
[97] Tolman, R. C. (1920), *J. Amer. Chem. Soc.*, **42**, 2506.
[98] Stern, M. J. and Weston, R. E. (1974), *J. Chem. Phys.*, **60**, 2803.
[99] Christov, S. G. (1965), *Ann. Phys.* (Leipzig), **15**, 87.
[100] Christov, S. G. (1968), *J. Res. Inst. Catalysis Hokkaido Univ.*, **16**, 169.
[101] Christov, S. G. (1972), *Ber. Bunsenges. Phys. Chem.*, **76**, 507.
[102] Bell, R. P. (1961), *Trans. Faraday Soc.*, **57**, 961.
[103] Bell, R. P. (1976), *J. Chem. Soc. Faraday II*, **72**, 2088.
[104] Bell, R. P. (1980), to be published.
[105] Bell, R. P., Sachs, W. H., and Tranter, R. L. (1971), *Trans. Faraday Soc.*, **67**, 1995.
[106] Kresge, A. J. (1973), *Chem. Soc. Reviews*, **2**, 475.
[107] Kresge, A. J. (1975), in *Proton Transfer Reactions*, ed. E. F. Caldin and V. Gold (Chapman and Hall, London) p. 179.
[108] Bell, R. P. (1978) in *Correlation Analysis in Chemistry, Recent Advances*, ed. N. B. Chapman and J. Shorter (Plenum Press, New York).
[109] Bell, R. P. (1935), *Proc. Roy. Soc. A*, **148**, 241.
[110] Weiss, J. J. (1964), *J. Chem. Phys.*, **41**, 1120.
[111] Bell, R. P. and Grainger, S. (1976), *J. Chem. Soc. Perkin II*, 1606.
[112] Melander, L. (1960), *Isotope Effects on Reaction Rates* (Ronald Press, New York).
[113] *Isotope Mass Effects in Chemistry and Biology* (Butterworths, London, 1964).
[114] *Pure Appl. Chem.* (1964), **8**, Nos. 3 and 4.
[115] Saunders, W. H. (1966), *Survey Progr. Chem.*, **3**, 109.
[116] Collins, C. J. and Bowman, N. S. (ed.) (1970), *Isotope Effects in Chemical Reactions* (Van Nostrand Reinhold, New York).
[117] Bell, R. P. (1974), *Chem. Soc. Reviews*, **3**, 513.
[118] Rock, P. A. (ed.) (1975), *Isotope Effects and Chemical Principles* (American Chemical Society, Washington D.C.).
[119] Soddy, F. (1933), *Proc. Roy. Soc. A*, **144**, 11.
[120] Urey, H. C. (1947), *J. Chem. Soc.*, 569.
[121] Bigeleisen, J. and Mayer, M. G. (1947), *J. Chem. Phys.*, **15**, 261.
[122] Zollinger, H. (1964), *Adv. Phys. Org. Chem.*, **2**, 253.
[123] Longuet-Higgins, H. C. (1955), *Phil. Mag.*, **46**, 98.
[124] Simpson, C. J. S. M. (1956), *J. Chem. Phys.*, **24**, 1108.
[125] Pearson, R. G. (1959), *J. Chem. Phys.*, **30**, 1537.
[126] Coté, G. L. and Thompson, H. W. (1951), *Proc. Roy. Soc. A*, **210**, 206.
[127] Van Hook, W. A. (1970), in *Isotope Effects in Chemical Reactions*, ed. C. J. Collins and N. S. Bowman (Von Nostrand Reinhold, New York) Tables 1.13 and 1.17.
[128] Shapiro, J. S. and Weston, R. E. (1972), *J. Phys. Chem.*, **76**, 1669.
[129] Sharp, T. E. and Johnston, H. S. (1962), *J. Chem. Phys.*, **37**, 1541.
[130] Kresge, A. J. and Chiang, Y. (1967), *J. Chem. Soc.*, B, 58.
[131] Kresge, A. J. and Chiang, Y. (1969), *J. Amer. Chem. Soc.*, **91**, 1025.
[132] Bigeleisen, J. (1949), *J. Chem. Phys.*, **17**, 675.

[133] Redlich, O. (1935), *Z. Phys. Chem. B*, **28**, 371.
[134] Wolfsberg, M. (1969), *Advances in Chemistry* (American Chemical Society, Washington D.C.) p. 89.
[135] Bron, J. and Wolfsberg, M. (1972), *J. Chem. Phys.*, **57**, 2862.
[136] Stern, M. J. and Wolfsberg, M. (1963), *J. Chem. Phys.*, **39**, 2776.
[137] Wolfsberg, M. and Stern M. J. (1964), *Pure Appl. Chem.*, **8**, 225, 325.
[138] Stern, M. J. and Wolfsberg, M. (1965), *J. Pharm. Sci.*, **54**, 849.
[139] Stern, M. J. and Wolfsberg, M. (1966), *J. Chem. Phys.*, **45**, 2618, 4105.
[140] Vogel, P. C. and Stern, M. J. (1971), *J. Chem. Phys.*, **54**, 779.
[141] Stern, M. J., Schneider, M. E., and Vogel, P. C. (1971), *J. Chem. Phys.*, **55**, 4286.
[142] Stern, M. J. and Vogel, P. C. (1971), *J. Amer. Chem. Soc.*, **93**, 4664.
[143] Schneider, M. E. and Stern, M. J. (1972), *J. Amer. Chem. Soc.*, **94**, 1517.
[144] Storm, B., Teklu, Y., and Sokolski, E. A. (1973), *Ann. N.Y. Acad. Sci.*, **206**, 631.
[145] Eaton, S. S. and Eaton, G. R. (1977), *J. Amer. Chem. Soc.*, **99**, 1601.
[146] Kreevoy, M. M. (1966), personal communication.
[147] Bell, R. P., Millington, J. P., and Pink, J. M. (1968), *Proc. Roy. Soc. A*, **303**, 1.
[148] Murrell, J. N. and Laidler, K. J. (1968), *Trans. Faraday Soc.*, **64**, 371.
[149] Swain, C. G., Stivers, E. C., Reuwer, J. F., and Schaad, L. J. (1958), *J. Amer. Chem. Soc.*, **80**, 5885.
[150] Bigeleisen, J. (1962), *Tritium in the Physical and Biological Sciences* (International Atomic Energy Agency, Vienna) Vol. 1, p. 161.
[151] More O'Ferrall, R. A. and Kouba, J. (1967), *J. Chem. Soc. B*, 985.
[152] Bell, R. P. (1973), *The Proton in Chemistry*, 2nd edn. (Chapman and Hall, London) Chapter 10.
[153] Reitz, O. (1936), *Z. Phys. Chem. A*, **176**, 363.
[154] Long, F. A. and Watson, D. (1958), *J. Chem. Soc.*, 2019.
[155] Bell, R. P. and Crooks, J. E. (1965), *Proc. Roy. Soc. A*, **286**, 285.
[156] Bell, R. P. and Tranter, R. L. (1974), *Proc. Roy. Soc. A*, **337**, 517.
[157] Bordwell, F. G. and Boyle, W. J. (1975), *J. Amer. Chem. Soc.*, **97**, 3447.
[158] Bell, R. P. and Goodall, D. M. (1966), *Proc. Roy. Soc. A*, **294**, 273.
[159] Barnes, D. J. and Bell, R. P. (1970), *Proc. Roy. Soc. A*, **318**, 421.
[160] Dixon, J. E. and Bruice, T. C. (1970), *J. Amer. Chem. Soc.*, **92**, 905.
[161] Cook, D., Hutchinson, R. E. J., Macleod, J. K., and Parker, A. J. (1974), *J. Org. Chem.*, **39**, 534.
[162] Keefe, J. R. and Munderloh, N. H. (1974), *J. Chem. Soc. Chem. Comm.*, 17.
[163] Hanna, S. B., Jermini, C., Loewenschuss, H., and Zollinger, H. (1974), *J. Amer. Chem. Soc.*, **96**, 7222.
[164] Jones, J. R. (1975), *Chem. Soc. Faraday Symp.*, **10**, 50.
[165] Anbar, M. and Meyerstein, D. (1964), *J. Phys. Chem.*, **68**, 3184.
[166] Bird, R. A. and Russell, K. E. (1965), *Canad. J. Chem.*, **43**, 2123.
[167] Pryor, W. A. and Kneipp, K. G. (1971), *J. Amer. Chem. Soc.*, **93**, 5584.
[168] Simonyi, M., Fitos, I., Kardos, J., Kovács, I., Lukovits, I., and Pospišil, J. (1977), *J. Chem. Soc. Faraday I*, **73**, 1286.
[169] Westheimer, F. H. (1961), *Chem. Rev.*, **61**, 265.
[170] Bigeleisen, J. (1964), *Pure Appl. Chem.* **8**, 217.
[171] Willi, A. V. and Wolfsberg, M. (1964), *Chem. and Ind.*, 2097.

# References

[172] Albery, W. J. (1967), *Trans. Faraday Soc.*, **63**, 200.
[173] Bell, R. P. (1966), *Disc. Faraday Soc.*, **39**, 16.
[174] Lewis, E. S. and Symons, M. C. R. (1958), *Quart. Rev. Chem. Soc.*, **12**, 230.
[175] Hawthorne, M. F. and Lewis, E. S. (1958), *J. Amer. Chem. Soc.*, **80**, 4296.
[176] Lewis, E. S. and Grinstein, R. H. (1962), *J. Amer. Chem. Soc.*, **84**, 1158.
[177] More O'Ferrall, R. A. (1970), *J. Chem. Soc. B*, 785.
[178] Stewart, R. (1976), in *Isotopes in Organic Chemistry*, ed. E. Buncel and C. C. Lee (Elsevier, Amsterdam) Vol. 2, Chapter 7.
[179] Willi, A. V. (1971), *Helv. Chim. Acta*, **54**, 1220.
[180] R. P. Bell, unpublished calculations.
[181] Katz, A. M. and Saunders, W. H. (1969), *J. Amer. Chem. Soc.*, **91**, 4469.
[182] Saunders, W. H. (1975), *Chem. Scr.*, **8**, 27.
[183] Saunders, W. H. (1976), *Chem. Scr.*, **10**, 82.
[184] Bunnett, J. F. (1962), *Angew. Chem. Int. Edit.*, **1**, 225.
[185] Cook, D., Hutchinson, R. E. J., Macleod, J. K., and Parker, A. J. (1974), *J. Org. Chem.*, **39**, 534.
[186] Cockerill, A. F. (1967), *J. Chem. Soc.*, B, 967.
[187] Bell, R. P. and Cox, B. G. (1970), *J. Chem. Soc. B*, 194.
[188] Bell, R. P. and Cox, B. G. (1971), *J. Chem. Soc. B*, 783.
[189] Earls, D. W., Jones, J. R., and Rumney, T. C. (1972), *J. Chem. Soc. Faraday I*, **68**, 925.
[190] Banger, J., Jaffe, A., Lin, A. -C., and Saunders, W. H. (1975), *Chem. Soc. Faraday Symp.*, **10**, 113.
[191] Blackwell, L. F. and Woodhead, J. L. (1975), *J. Chem. Soc. Perkin II*, 234.
[192] Bowden, K. (1966), *Chem. Rev.*, **66**, 119.
[193] Cox, B. G. and Gibson, A. (1977), *J. Chem. Soc. Perkin II*, 1812.
[194] Blackwell, L. F. and Woodhead, J. L. (1975), *J. Chem. Soc. Perkin II*, 1218.
[195] Caldin, E. F. and Mateo, S. (1973), *J. Chem. Soc. Chem. Comm.*, 854.
[196] Caldin, E. F. and Mateo, S. (1975), *J. Chem. Soc. Faraday I*, **71**, 1876.
[197] Marcus, R. A. (1968), *J. Phys. Chem.*, **72**, 891.
[198] Kreevoy, M. M. and Oh, S. (1973), *J. Amer. Chem. Soc.*, **95**, 4805.
[199] Kreevoy, M. M. (1976) in *Isotopes in Organic Chemistry*, ed. E. Buncel and C. C. Lee (Elsevier, Amsterdam).
[200] Hassid, A. I., Kreevoy, M. M., and Liang, T.-M. (1975), *Chem. Soc. Faraday Symp.*, **10**, 69.
[201] Kresge, A. J., Sagatys, D. S., and Chen, H. -L. (1977), *J. Amer. Chem. Soc.*, **99**, 7228.
[202] Bell, R. P., (1978), *Finnish Chem. Lett.*, 7.
[203] Band, Y. B. and Freed, K. F. (1975), *J. Chem. Phys.*, **63**, 4479.
[204] Stern, M. J. and Weston, R. E. (1974), *J. Chem. Phys.*, **60**, 2808.
[205] Lewis, E. S. and Robinson, J. K. (1968), *J. Amer. Chem. Soc.*, **90**, 4337.
[206] Stern, M. J. and Weston, R. E. (1974), *J. Chem. Phys.*, **60**, 2815.
[207] Jones, J. R. (1969), *Trans. Faraday Soc.*, **65**, 2138.
[208] Hine, J., Kaufman, J. C., and Cholod, M. S. (1972), *J. Amer. Chem. Soc.*, **94**, 4590.
[209] Melander, L. and Bergman, N. (1976), *Acta Chem. Scand. A* **30**, 703.

[210] German, E. D. (1977), *Izvest. Akad. Nauk S.S.S.R, Ser. Khim.*, 2802.
[211] Lewis, E. S., Perry, J. M., and Grinstein, R. H. (1970), *J. Amer. Chem. Soc.*, **92** 899.
[212] Isaacs, N. S., Javaid, K., and Rannala, E. (1978), *J. Chem. Soc. Perkin II*, 709.
[213] Palmer, D. A. and Kelm, H. (1977), *Aust. J. Chem.*, **30**, 1229.
[214] Sicking, G. (1972), *Ber. Bunsenges. Phys. Chem.*, **76**, 790.
[215] De Ribaupierre, Y. and Manchester, F. D. (1973), *J. Phys. C*, **6**, L390.
[216] Guil, J. M., Hayward, D. O., and Taylor, N. (1973), *Proc. Roy. Soc. A*, **335**, 141.
[217] Simonyi, M. and Mayer, I. (1975), *J. Chem. Soc. Chem. Comm.*, 695.
[218] Bell, R. P., Fendley, J. A. and Hulett, J. R. (1956), *Proc. Roy. Soc. A*, **235**, 453.
[219] Sprague, E. D. and Williams, F. (1971), *J. Amer. Chem. Soc.*, **93**, 787.
[220] Le Roy, R. J., Sprague, E. D., and Williams, F. (1972), *J. Phys. Chem.*, **76**, 546.
[221] Wang, J. T. and Williams, F. (1972), . *Amer. Chem. Soc.*, **94**, 2930.
[222] Campion, A. and Williams, F. (1972), *J. Amer. Chem. Soc.*, **94**, 7633.
[223] Hudson, R. L., Shiotani, M., and Williams, F. (1977), *Chem. Phys. Lett.*, **48**, 193.
[224] Sprague, E. D. (1977), *J. Phys. Chem.*, **81**, 516.
[225] Bonin, M. A., Tsuji, K., and Williams, F. (1968), *Nature*, **218**, 946.
[226] Egland, R. J. and Symons, M. C. R. (1970), *J. Chem. Soc. A*, 1326.
[227] Wijnen, M. H. J. (1954), *J. Chem. Phys.*, **22**, 1074.
[228] Milligan, D. E. and Jacox, M. E. (1962), *J. Mol. Spectrosc.*, **8**, 126.
[229] Pace, E. L. and Noe, L. J. (1968), *J. Chem. Phys.*, **49**, 5317.
[230] Trotman-Dickenson, A. F. and Steacie, E. W. R. (1951), *J. Chem. Phys.*, **19**, 329.
[231] Barclay, L. R. C., Griller, D., and Ingold, K. U. (1974), *J. Amer. Chem. Soc.*, **96**, 3011.
[232] Brunton, G., Griller, D., Barclay, L. R. C., and Ingold, K. U. (1976), *J. Amer. Chem. Soc.*, **98**, 6803.
[233] Brunton, G., Gray, J. A., Griller, D., Barclay, L. R. C., and Ingold, K. U. (1978), *J. Amer. Chem. Soc.*, **100**, 4197.
[234] Yakimchenko, O. E. and Lebedev, Ya. S. (1971), *Int. J. Radiat. Phys. Chem.*, **3**, 17.
[235] Toryiama, K., Nunome, K., and Iwasaki, M. (1977), *J. Amer. Chem. Soc.*, **99**, 5823.
[236] Willard, J. E. (1968), in *Fundamental Processes in Radiation Chemistry*, ed. P. Ausloos (Wiley – Interscience, New York) p. 599.
[237] French, W. G. and Willard, J. E. (1968), *J. Phys. Chem.*, **72**, 4604.
[238] Neiss, M. A. and Willard, J. E. (1975), *J. Phys. Chem.*, **79**, 783.
[239] Neiss, M. A., Sprague, E. D. and Willard, J. E. (1975), *J. Chem. Phys.*, **63**, 1118.
[240] Wilkey, D. D. and Willard, J. E. (1976), *J. Chem. Phys.*, **64**, 3976.
[241] Dubinskaya, A. M. and Butyagin, P. Yu. (1973), *Dokl. Akad. Nauk S.S.S.R.*, **211**, 143.
[242] Bromberg, A., Muszkat, K. A., and Fischer, E. (1968), *Chem. Comm.*, 1352.
[243] Bromberg, A., Muszkat, K. A., Fischer, E., and Klein, F. S. (1972), *J. Chem. Soc. Perkin II*, 588.
[244] Warshel, A. and Bromberg, A. (1970), *J. Chem. Phys.*, **52**, 1262.
[245] Bromberg, A., Muszkat, K.A., and Warshel, A. (1970), *J. Chem. Phys.*, **52**, 5952.
[246] Haven, Y., Williams, R. C., Hamrick, P. J., and Shields, J. (1974), *J. Chem. Phys.*, **60**, 127.

# References

[247] London, F. (1929), *Z. Electrochem.*, **35**, 552.
[248] Johnston, H. S. (1966), *Gas Phase Reaction Rate Theory* (Ronald Press, New York) Chapter 10.
[249] Weston, R. E. (1967), *Science*, **158**, 332.
[250] Van Hook, W. A. (1970), in Reference 113, pp. 63–73.
[251] Truhlar, D. G. and Wyatt, R. E. (1976), *Ann. Rev. Phys. Chem.*, **27**, 1.
[252] Truhlar, D. G. and Wyatt, R. E. (1977), *Adv. Chem. Phys.*, **27**, 50.
[253] Farkas, A. (1930), *Z. Electrochem.*, **36**, 782.
[254] Farkas, A. (1930), *Z. Phys. Chem. B*, **10**, 419.
[255] Geib, K. H. and Harteck, P. (1931), *Z. Phys. Chem. Bodenstein Festb.*, 849.
[256] Farkas, A. and Farkas, L. (1935), *Proc. Roy. Soc. A*, **152**, 124.
[257] Boato, G., Careri, G., Cimino, A., Molinari, E., and Volpi, G. G. (1956), *J. Chem. Phys.*, **24**, 783.
[258] Schulz, W. R. and Le Roy, D. J. (1964), *Canad. J. Chem.*, **42**, 2480.
[259] Ridley, B. A., Schulz, W. R., and Le Roy, D. J. (1966), *J. Chem. Phys.*, **44**, 3344.
[260] Westenburg, A. A. and De Haas, N. (1967), *J. Chem. Phys.*, **47**, 1393.
[261] Mitchell, D. N. and Le Roy, D. J. (1973), *J. Chem. Phys.*, **58**, 3449.
[262] Schulz, W. R. and Le Roy, D. J. (1965), *J. Chem. Phys.*, **42**, 3869.
[263] Le Roy, D. J., Ridley, B. A., and Quickert, K.A. (1967), *Disc. Faraday Soc.*, **44**, 97.
[264] Niki, H. and Mains, G. J. (1972), *J. Phys. Chem.*, **76**, 3538.
[265] Shavitt, I., Stevens, R. M., Minn, F. L., and Karplus, M. (1968), *J. Chem. Phys.*, **48**, 2700.
[266] Shavitt, I. (1968), *J. Chem. Phys.*, **49**, 4048.
[267] Quickert, K. A. and Le Roy, D. J. (1970), *J. Chem. Phys.*, **53**, 1325.
[268] Quickert, K. A. and Le Roy, D. J. (1971), *J. Chem. Phys.*, **54**, 5444.
[269] Liu, B. (1973), *J. Chem. Phys.*, **58**, 1925.
[270] Yates, A. C. and Lester, W. A. (1974), *Chem. Phys. Lett.*, **24**, 305.
[271] Koeppl, G. W. (1973), *J. Chem. Phys.*, **59**, 3425.
[272] Marcus, R. A. and Coltrin, M. E. (1977), *J. Chem. Phys.*, **67**, 2609.
[273] Schatz, G. C. and Kuppermann, A. (1976), *J. Chem. Phys.*, **65**, 4642,4668.
[274] Connor, J. N. L. (1974), *Chem. Soc. Ann. Rep. A*, **70**, 5.
[275] Porter, R. N. and Karplus, M. (1964), *J. Chem. Phys.*, **40**, 1105.
[276] Mortensen, E. M. and Pitzer, K. S. (1962), *Chem. Soc. Spec. Pub.*, **16**, 57.
[277] Child, M. S. (1967), *Disc. Faraday Soc.*, **44**, 68.
[278] Miller, W. H. (1974), *Adv. Chem. Phys.*, **25**, 69.
[279] Doll, J. D., George, T. F., and Miller, W. H. (1973), *J. Chem. Phys.*, **58**, 1343.
[280] Miller, W. H. (1974), *J. Chem. Phys.*, **61**, 1823.
[281] Miller, W. H. (1975), *J. Chem. Phys.*, **62**, 1899.
[282] Miller, W. H. (1975), *J. Chem. Phys.*, **63**, 1166.
[283] Rollefson, G. K. (1934), *J. Chem. Phys.*, **2**, 144.
[284] Bigeleisen, J., Klein, F. S., Weston, R. E., and Wolfsberg, M. (1959), *J. Chem. Phys.*, **30**, 1340.
[285] Chiltz, G., Eckling, R., Goldfinger, P., Huybrechts, G., Johnston, H. S., Meyers, L., and Verbekel, G. (1963), *J. Chem. Phys.*, **38**, 1053.
[286] Persky, A. and Klein, F. S. (1966), *J. Chem. Phys.*, **44**, 3617.
[287] Bar Yaakov, Y., Persky, A., and Klein, F. S. (1973), *J. Chem. Phys.*, **59**, 2415.

[288] Stern, M. J., Persky, A., and Klein, F. S. (1973), *J. Chem. Phys.*, **58**, 5697.
[289] Sato, S. (1955), *J. Chem. Phys.*, **23**, 592, 2465.
[290] Johnston, H. S. and Parr, C. (1963), *J. Amer. Chem. Soc.*, **85**, 2544.
[291] Ashmore, P. G. and Chanmugan, J. (1953), *Trans. Faraday Soc.*, **49**, 254.
[292] Persky, A. and Baer, M. (1974), *J. Chem. Phys.*, **60**, 133.
[293] Baer, M. (1974), *Mol. Phys.*, **27**, 1429.
[294] Baer, M., Halavee, U., and Persky, A. (1974), *J. Chem. Phys.*, **61**, 5122.
[295] Persky, A. (1973), *J. Chem. Phys.*, **59**, 3612, 5578.
[296] Persky, A. (1974), *J. Chem. Phys.*, **60**, 3354.
[297] Koeppl, G. W. (1974), *J. Chem. Phys.*, **60**, 1684.
[298] Schatz, G. C., Bowman, J. M. and Kuppermann, A. (1975), *J. Chem. Phys.*, **63**, 674, 685.
[299] Johnston, H. S. and Tsuikow-Roux, E. (1962), *J. Chem. Phys.*, **36**, 463.
[300] Kurylo, M. J., Hollinden, G. A., and Timmons, R. S.(1970), *J. Chem. Phys.*, **52**, 1773.
[301] Clark, T. C. and Dove, J. E. (1973), *Canad. J. Chem.*, **51**, 2147, 2155.
[302] Pacey, P. D. (1973), *Canad. J. Chem.*, **51**, 2146.
[303] McNesby, J. R. (1960), *J. Phys. Chem.*, **64**, 1671.
[304] Arthur, N. L., Donchi, K. F., and McDonell, J. A. (1975), *J. Chem. Soc. Faraday I*, **71**, 2431, 2442.
[305] Smith, I. W. M. and Zellner, R. (1974), *J. Chem. Soc. Faraday II*, **70**, 1045.
[306] Ahlberg, P. (1973), *Chem. Scr.*, **4**, 33.
[307] Ahlberg, P. and Bengtsson, S. (1974), *Chem. Scr.*, **6**, 45.
[308] Thibblin, A. and Ahlberg, P. (1976), *Acta Chem. Scand. B*, **30**, 555.
[309] Thibblin, A., Bengtsson, S., and Ahlberg, P. (1977), *J. Chem. Soc. Perkin II*, 1569.
[310] Thibblin, A. and Ahlberg, P. (1977), *J. Amer. Chem. Soc.*, **99**, 7926.
[311] Krumbiegel, P. (1968), *Z. Chem.*, **8**, 328.
[312] Rummel, S. and Huebner, H. (1969), *Z. Chem.*, **9**, 150.
[313] Jones, J. R. (1969), *Trans. Faraday Soc.*, **65**, 2430.
[314] Jones, J. R., Marks, R. E., and Subba Rao, S. C. (1967), *Trans. Faraday Soc.*, **63**, 993.
[315] Wong, S. M., Fischer, H. F., and Cram, D. J. (1971), *J. Amer. Chem. Soc.*, **93**, 2235.
[316] Lewis, E. S. and Allen, J. D. (1964), *J. Amer. Chem. Soc.*, **86**, 2022.
[317] Funderburk, L. and Lewis, E. S. (1964), *J. Amer. Chem. Soc.*, **86**, 2531.
[318] Lewis, E. S. and Funderburk, L. (1967), *J. Amer. Chem. Soc.*, **89**, 2322.
[319] Lewis, E. S. and Robinson, J. K. (1968), *J. Amer. Chem. Soc.*, **90**, 4337.
[320] Wilson, H., Caldwell, J. D., and Lewis, E. S. (1973), *J. Org. Chem.*, **38**, 564.
[321] Caldin, E. F. and Tomalin, G. (1968), *Trans. Faraday Soc.*, **64**, 2814.
[322] Caldin, E. F., Jarczewski, A., and Leffek, K. T. (1971), *Trans. Faraday Soc.*, **67**, 110.
[323] Caldin, E. F. and Mateo, S. (1976), *J. Chem. Soc. Faraday I*, **72**, 112.
[324] Caldin, E. F. and Wilson, C. J. (1975), *Chem. Soc. Faraday Symp.*, **10**, 121.
[325] Caldin, E. F., Parboo, D. M., Walker, F. A., and Wilson, C. J. (1976), *J. Chem. Soc. Faraday I*, **72**, 1856.

# References

[326] Caldin, E. F., Parboo, D. M., and Wilson, C. J. (1976), *J. Chem. Soc. Faraday I*, **72**, 2645.
[327] Shiner, V. J. and Smith, M. L. (1961), *J. Amer. Chem. Soc.*, **83**, 593.
[328] Shiner, V. J. and Martin, B. (1964), *Pure Appl. Chem.*, **8**, 371.
[329] Bethell, D. and Cockerill, A. F. (1966), *J. Chem. Soc.*, B, 917.
[330] Blackwell, L. F., Buckley, P. D., Jolley, K. W., and McGibbon, A. K. H. (1973), *J. Chem. Soc. Perkin II*, 169.
[331] Lifschitz, G. and Stein, G. (1962), *J. Chem. Soc.*, 3706.
[332] Vacek, K. and von Sonntag, C. (1969), *J. Chem. Soc. Chem. Comm.*, 1256.
[333] Antonovskii, V. L. and Berezin, I. V. (1960), *Zhur. fiz. Khim.*, **34**, 1286.
[334] Lewis, E. S. and Butler, M. M. (1971), *J. Org. Chem.*, **36**, 2582.
[335] Shishkina, L. N. and Berezin, I. V. (1965), *Zhur. fiz. Khim.*, **39**, 2547.
[336] Larson, G. F. and Gilliom, R. D. (1975), *J. Amer. Chem. Soc.*, **97**, 3444.
[337] Roberts, R. J. and Ingold, K. U. (1973), *J. Amer. Chem. Soc.*, **95**, 3228.
[338] Simonyi, M. and Tüdös, F. (1970), *Adv. Phys. Org. Chem.*, **9**, 127.
[339] Roth, W. R. and Konig, J. (1966), *Ann. Chem.*, **699**, 24.
[340] McLean, S., Webster, C. J., and Rutherford, R. J. D. (1969), *Canad. J. Chem.*, **47**, 1555.
[341] Bowman, D. F., Gillan, T., and Ingold, K. U. (1971), *J. Amer. Chem. Soc.*, **93**, 6555.
[342] Stewart, R. and van der Linden, R. (1960), *Disc. Faraday Soc.*, **29**, 211.
[343] Stewart, R. and Mocek, M. M. (1963), *Canad. J. Chem.*, **41**, 1161.
[344] Stewart, R. and Lee, D. G. (1964), *Canad. J. Chem.*, **42**, 439.
[345] Goldman, I. M. (1969), *J. Org. Chem.*, **34**, 3489.
[346] Rogne, O. (1977), *J. Chem. Soc. Chem. Comm.*, 695.
[347] Blanch, J. H. and Rogne, O. (1978), *J. Chem. Soc. Faraday I*, **74**, 1263.
[348] Rogne, O. (1978), *Acta Chem. Scand. A*, **32**, 559.
[349] Caldin, E. F. (1979), Abstracts, 2nd *Internat. Conf. on Reaction Mechanisms in Solution* (Chem. Soc. 1979), p. 161.
[350] Kim, J. H. and Leffek, K. T. (1974), *Canad. J. Chem.*, **52**, 592.
[351] Jarczewski, A., Pruszynski, P., and Leffek, K. T. (1975), *Canad. J. Chem.*, **53**, 1176.
[352] Bell, R. P. and Onwood, D. P. (1967), *J. Chem. Soc. B*, 150.
[353] Lewis, E. S., Allen, J. D., and Wallick, E. T. (1969), *J. Org. Chem.*, **34**, 255.
[354] Willi, A. V. (1966), *J. Phys. Chem.*, **70**, 2705.
[355] Caldin, E. F. (1969), *Chem. Rev.*, **69**, 135.
[356] Caldin, E. F. and Harbron, E. (1962), *J. Chem. Soc.*, 3454.
[357] Caldin, E. F. and Jackson, R. A. (1960), *J. Chem. Soc.*, 2410.
[358] Caldin, E. F. and Kasparian, M. (1965), *Disc. Faraday Soc.*, **39**, 25.
[359] Kurz, J. L. and Kurz, L. C. (1972), *J. Amer. Chem. Soc.*, **94**, 4451.
[360] Goldanskii, V. I. (1975), *Uspekhi Khim.*, **44**, 2121 (English translation., *Russian Chem. Rev.*, **44**, 1019).
[361] Goldanskii, V. I. (1976), *Ann. Rev. Phys. Chem.*, **27**, 85.
[362] Kiryukhin, D. P., Kaplan, A. M., Barkalov, I. M., and Goldanskii, V. I. (1970), *Vysokomol. Soedin*, **12** (B), 491.
[363] Kiryukhin, D. P., Kaplan, A. M., Barkalov, I. M., and Goldanskii, V. I. (1971), *Dokl. Akad. Nauk S.S.S.R.*, **199**, 857.

[364] Kiryukhin, D. P., Kaplan, A. M., Barkalov, I. M., and Goldanskii, V. I. (1972), *Vysokomol. Soedin.*, **14 (A)**, 2115.
[365] Kiryukhin, D. P., Kaplan, A. M., Barkalov, I. M., and Goldanskii, V. I. (1972), *Dokl. Akad. Nauk S.S.S.R.*, **206**, 147.
[366] Goldanskii, V. I., Frank-Kamenetskii, M. D., and Barkalov, I. M. (1973), *Dokl. Akad. Nauk S.S.S.R.*, **211**, 133.
[367] Goldanskii, V. I., Frank-Kamenetskii, M. D., and Barkalov, I. M. (1973), *Science*, **182**, 1344.
[368] Kiryukhin, D. P., Barkalov, I. M., and Goldanskii, V. I. (1974), *Vysokomol. Soedin.*, **16 (B)**, 565.
[369] Goldanskii, V. I., Kiryukhin, D. P., and Barkalov, I. M. (1974), *Khim. Vys. Energ.*, **8**, 279.
[370] Austin, R. H., Beeson, K. W., Chan, S. S., Debrunner, P. G., Downing, R., Eisenstein, L., Frauenfelder, H., and Nordlund, T. M. (1976), *Rev. Sci. Instrum.*, **47**, 445.
[371] Austin, R. H., Beeson, K. W., Eisenstein, L., Frauenfelder, H., Gunsalus, I. C., and Marshall, V. P. (1974), *Phys. Rev. Lett.*, **32**, 403.
[372] Austin, R. H., Beeson, K. W., Eisenstein, L., Frauenfelder, H., and Gunsalus, I. C. (1975), *Biochemistry*, **14**, 5355.
[373] Alberding, N., Austin, R. H., Beeson, K. W., Chan, S. S., Eisenstein, L., Frauenfelder, H., and Nordlund, T. M. (1976), *Science*, **192**, 1002.
[374] Primack, W. (1955), *Phys. Rev.*, **100**, 1677.
[375] Primack, W. (1960), *J. Appl. Phys.*, **31**, 1525.
[376] Zamaraev, K. I., Khairutdinov, R. F., Mikhailov, A. I., and Goldanskii, V. I. (1971), *Dokl. Akad. Nauk S.S.S.R.*, **199**, 640.
[377] Mikhailov, A. I., Kuzina, S. I., Lukovnikov, A. F., and Goldanskii, V. I. (1972), *Dokl. Akad. Nauk S.S.S.R.*, **204**, 383.
[378] Mikhailov, A. I., Bolshakov, A. I., Lebedev, Ya. S., and Goldanskii, V. I. (1972), *Fiz. Tverd. Tela*, **14**, 1172.
[379] Alberding, N., Austin, R. H., Chan, S. S., Eisenstein, L., Frauenfelder, H., Gunsalus, I. C., and Nordlund, T. M. (1976), *J. Chem. Phys.*, **65**, 4701.
[380] Brewer, J. H., Crowe, K. M., Gygax, F. N., Johnson, R. F., Fleming, D. G., and Schenk, A. (1974), *Phys. Rev. A*, **9**, 495.
[381] Fleming, D. G., Brewer, J. H., Garner, D. M., Pifer, A. E., Bowen, T., Delise, D. A., and Crowe, K. M. (1976), *J. Chem. Phys.*, **64**, 4.
[382] Herzberg, G., (1950), *Molecular Spectra and Molecular Structure. I. Diatomic Molecules*, 2nd edn. (Van Nostrand, Princeton, N.J.) pp. 405–450.
[383] Herzberg, G. (1971), *Spectra and Structures of Simple Free Radicals* (Cornell U. P., Ithaca, N. Y.) Chapter 5.
[384] Mulliken, R. S. (1960), *J. Chem. Phys.*, **33**, 247.
[385] Olsson, E. (1938), Thesis, Stockholm.
[386] Way, K. R. and Stwalley, W. C. (1973), *J. Chem. Phys.*, **59**, 5298.
[387] Stwalley, W. C. (1975), *J. Chem. Phys.*, **63**, 3062.
[388] Johns, J. W. C. (1963), *Canad. J. Phys.*, **41**, 209.
[389] Douglas, A. E. (1963), *Disc. Faraday Soc.*, **35**, 158.
[390] Herzberg, G. (1961), *Proc. Roy. Soc. A*, **262**, 291.

[391] Robinson, G. W. (1961), *J. Mol. Spectr.*, **6**, 58.
[392] McCoy, E. F. and Ross, I. G. (1962), *Aust. J. Chem.*, **15**, 573.
[393] Hunt, G. R., McCoy, E. F., and Ross, I. G. (1962), *Aust. J. Chem.*, **15**, 591.
[394] Beddard, G. S., Fleming, G. R., Gijzeman, O. L. J., and Porter, G. (1973), *Chem. Phys. Lett.*, **18**, 481.
[395] Formosinho, S. J. (1974), *J. Chem. Soc. Faraday II*, **70**, 605.
[396] Herzberg, G. (1966), *Molecular Spectra and Molecular Structure. II. Infrared and Raman Spectra of Polyatomic Molecules* (Van Nostrand, New York) p. 221.
[397] Townes, C. H. and Schawlow, A. L. (1955), *Microwave Spectroscopy* (McGraw-Hill, New York) Chapter 12.
[398] Sugden, T. M. and Kenney, C. N. (1965), *Microwave Spectroscopy of Gases* (Van Nostrand, London) Chapter 8.
[399] Cleeton, C. E. and Williams, N. H. (1934), *Phys. Rev.*, **45**, 234.
[400] Barrow, G. M. (1960), *Spectrochim. Acta*, **16**, 799.
[401] Nakai, Y. and Hirota, K. (1959), *Bull. Chem. Soc. Japan*, **32**, 769.
[402] Costain, C. C. and Srivastava, G. P. (1964), *J. Chem. Phys.*, **41**, 1620.
[403] Excoffon, P. and Maréchal, Y. (1972), *Spectrochim. Acta. A*, **28**, 269.
[404] Ginn, S. W. G. and Wood, J. L. (1967), *J. Chem. Phys.*, **46**, 2735.
[405] Rothschild, W. G. (1974), *J. Chem. Phys.*, **61**, 3422.
[406] Hadzi, D. (1961), *J. Chem. Phys.*, **34**, 1445.
[407] Haas, C. and Hornig, D. F. (1960), *J. Chem. Phys.*, **32**, 1763.
[408] Kampschultesche, I. and Zundel, G. (1970), *J. Phys. Chem.*, **74**, 2363.
[409] Lindemann, R. and Zundel, G. (1972), *J. Chem. Soc. Faraday II.* **68**, 979.
[410] Brickmann, J. and Zimmermann, H. (1966), *Ber. Bunsenges. Phys. Chem.*, **70**, 521.
[411] Brickmann, J. and Zimmermann, H. (1967), *Ber. Bunsenges. Phys. Chem.*, **71**, 160.
[412] Brickmann, J. and Zimmermann, H. (1968), *Z. Naturforschung.*, **23a**, 1.
[413] Brickmann, J. (1971), *Ber. Bunsenges. Phys. Chem.*, **75**, 747.
[414] Lin, C. C. and Swalen, J. D. (1959), *Rev. Mod. Phys.*, **31**, 841.
[415] Srinivasan, R. (1975), *Int. Rev. Phys. Chem.*, ed. C. A. McDowell Series 2, Vol. 4 (Butterworths, London), pp. 209-226.
[416] Thomas, J. T., Alpert, N. L., and Torrey, H. C. (1950), *J. Chem. Phys.*, **18**, 1511.
[417] Tomita, K. (1953), *Phys. Rev.*, **89**, 429.
[418] Andrew, E. R. and Bersohn, R. (1950), *J. Chem. Phys.*, **18**, 159.
[419] Carolan, J. L. and Scott, T. A. (1970), *J. Magn. Resonance*, **2**, 243.
[420] Apaydin, F. and Clough, S. (1968), *J. Phys.*, C, **1**, 932.
[421] Ripmeester, J. A., Garg, S. K., and Davidson, D. W. (1977), *J. Chem. Phys.*, **67**, 2275.
[422] Hecht, K. T. and Dennison, D. M. (1957), *J. Chem. Phys.*, **26**, 31.
[423] Clough, S., Hobson, T., and Nugent, S. M. (1975), *J. Phys.*, C, **8**, L95.
[424] Johnson, C. S. and Mottley, C. (1973), *Chem. Phys. Lett.*, **22**, 430.
[425] Bersohn, R. and Gutowsky, H. S. (1954), *J. Chem. Phys.*, 651.
[426] Watton, A., Sharp, A. R., Petch, H. E., and Pintar, M. M. (1972), *Phys. Rev. B*, **5**, 4281.
[427] Dunn, M. B., Ikeda, R., and McDowell, C. A. (1972), *Chem. Phys. Lett.*, **16**, 226.

[428] Ikeda, R. and McDowell, C. A. (1973), *Mol. Phys.*, **25**, 1217.
[429] Smith, D. (1978), *J. Chem. Phys.*, **68**, 619.
[430] Kanzig, W. (1962), *J. Phys. Chem. Solids*, **23**, 479.
[431] Freed, J. H. (1965), *J. Chem. Phys.*, **43**, 1710.
[432] Davidson, R. B. and Miyagawa, I. (1970), *J. Chem. Phys.*, **52**, 1727.
[433] Roberts, J. D. (1959), *Nuclear Magnetic Resonance: Applications to Organic Chemistry* (McGraw–Hill, New York) Chapter 4.
[434] Pople, J. A., Schneider, W. G., and Bernstein, H. J. (1959), *High-Resolution Nuclear Magnetic Resonance* (McGraw–Hill, New York) Chapter 10.
[435] Meiboom, S. (1960), *Z. Electrochem.*, **64**, 50.
[436] Loewenstein, A. (1963), in *Fluctuation, Relaxation and Resonance in Magnetic Systems*, ed. D. ter Haar (Oliver and Boyd, Edinburgh and London) pp. 261–268.
[437] Loewenstein, A. and Connor, T. M. (1963), *Ber. Bunsenges. Phys. Chem.*, **67**, 280.
[438] Strehlow, H. (1963), in *Technique of Organic Chemistry*, ed. E.L. Friess, E. S. Lewis and A. Weissberger, Vol. 8, Part II (Interscience, New York), Chapter 17.
[439] Caldin, E. F. (1964), *Fast Reactions in Solution* (Blackwell, Oxford) Chapter 11.
[440] Emsley, J. W., Feeney, J., and Sutcliffe, L. H. (1965), *High Resolution Nuclear Magnetic Resonance Spectroscopy* (Pergamon, Oxford).
[441] Heidberg, J. (1968), in *Chemische Elementarprozesse*, ed. H. Hartmann (Springer, Berlin).
[442] Carter, R. E., Drakenberg, T., and Bergman, N. (1975), *J. Amer. Chem. Soc.*, **97**, 6990.
[443] Mitchell, R. W., Burr, J. C., and Merritt, J. A. (1967), *Spectrochim. Acta A*, **23**, 195.
[444] Abraham, R. J., Hawkes, G. E., and Smith, K. M. (1974), *J. Chem. Soc. Perkin II*, 627.
[445] Abraham, R. J., Hawkes, G. E., and Smith, K. M. (1974), *Tetrahed. Lett.*, 1483.
[446] Gust, D. and Roberts, J. D. (1977), *J. Amer. Chem. Soc.*, **99**, 3637.
[447] Bell, R. P. and Critchlow, J. E. (1971), *Proc. Roy. Soc. A*, **325**, 35.
[448] Hennig, J. and Limbach, H. (1979), *J. Chem. Soc. Faraday II*, **75**, 752; *J. Chem. Phys.*, **71**, 3120.
[449] Abraham, R. J., Hawkes, G. E., and Smith, K. M. (1974), *Tetrahed. Lett.*, 71.
[450] Furhop, J. H. (1976), *Angew. Chem.*, **19**, 704.
[451] Brickmann, J. (1971), *Ber. Bunsenges. Phys. Chem.*, **75**, 747.
[452] Mason, S. (1958), *J. Chem. Soc.*, 976.
[453] Silvers, S. J. and Tulinsky, A. (1967), *J. Amer. Chem. Soc.*, **89**, 3331.
[454] Webb, L. E. and Fleischer, E. B. (1965), *J. Chem. Phys.*, **43**, 3100.
[455] Tulinsky, A. (1973), *Ann. N.Y. Acad. Sci.*, **206**, 47.
[456] Hamor, M. J., Hamor, T. A., and Hoard, J. L. (1964), *J. Amer. Chem. Soc.*, **86**, 1938.
[457] Lutz, M. (1974), *Spectrosc. Lett.*, **7**, 133.
[458] Van Dorp, W. G., Soma, M. T. J., Kooter, J. A., and van der Waals, J. H. (1973), *Chem. Phys. Lett.*, **37**, 443.
[459] Van Dorp, W. G., Soma, M. T. J., Kooter, J. A., and van der Waals, J. H. (1974), *Mol. Phys.*, **28**, 1551.

# References

[460] Muetterties, E. L. (1970), *Acc. Chem. Res.*, **3**, 266.
[461] Gutowsky, H. S., McCall, A. D., and Schlichter, C. P. (1953), *J. Chem. Phys.*, **21**, 279.
[462] Berry, R. S. (1960), *J. Chem. Phys.*, **32**, 936.
[463] Holmes, R. R. (1968), *Inorg. Chem.*, **7**, 2229.
[464] Muetterties, E. L. and Phillips, W. D. (1957), *J. Amer. Chem. Soc.*, **79**, 322.
[465] Muetterties, E. L. and Phillips, W. D. (1959), *J. Amer. Chem. Soc.*, **81**, 1084.
[466] Downs, J. J. and Johnson, R. E. (1954), *J. Chem. Phys.*, **22**, 143.
[467] Downs, J. J. and Johnson, R. E. (1955), . *Amer. Chem. Soc.*, **77**, 2098.
[468] Christov, S. G. (1945), *Ann. Univ. Sofia, Fac. Phys. Math.*, **42** (2), 69.
[469] Christov, S. G. (1946), *Ann. Univ. Sofia, Fac. Phys. Math.*, **43** (2), 63.
[470] Christov, S. G. (1948), *C.R. Acad. Bulg. Sci.*, **1**, 43.
[471] Christov, S. G. (1958), *Z. Electrochem.*, **62**, 567.
[472] Christov, S. G. (1959), *Dokl. Akad. Nauk S.S.S.R.*, **125**, 141.
[473] Christov, S. G. (1960), *Dokl. Akad. Nauk S.S.S.R.*, **136**, 663.
[474] Christov, S. G. (1960), *Z. Phys. Chem. (Leipzig)*, **214**, 40.
[475] Christov, S. G. (1962), *Electrochim. Acta*, **4**, 194, 306.
[476] Christov, S. G. (1964), *Electrochim. Acta*, **9**, 575.
[477] Christov, S. G. (1965), *Disc. Faraday Soc.*, **39**, 60, 254, 263.
[478] Christov, S. G. (1968), *J. Res. Inst. Catal. Hokkaido Univ.*, **16**, 169.
[479] Christov, S. G. (1972), *Croat. Chem. Acta*, **44**, 67.
[480] Conway, B. E. (1959), *Canad. J. Chem.*, **37**, 178.
[481] Conway, B. E. and Salomon, M. (1964), *J. Chem. Phys.*, **41**, 3169.
[482] Conway, B. E. and Salomon, M. (1964), *J. Phys. Chem.*, **68**, 2009.
[483] Conway, B. E. and Salomon, M. (1964), *Ber. Bunsenges. Phys. Chem.*, **68**, 231.
[484] Conway, B. E. and Salomon, M. (1965), *Disc. Faraday Soc.*, **39**, 223, 261.
[485] Salomon, M., Enke, C. G., and Conway, B. E. (1965), *J. Chem. Phys.*, **43**, 3986.
[486] Conway, B. E., Mackinnon, D. J., and Tilak, B. V. (1970), *Trans. Faraday Soc.*, **66**, 1203.
[487] Bockris, J. O'M. and Srinivasan, S. (1964), *J. Electrochem. Soc.*, **111**, 844.
[488] Bockris, J. O'M. and Srinivasan, S. (1964), *Electrochim. Acta*, **9**, 31.
[489] Bockris, J. O'M., Srinivasan, S., and Matthews, D. B. (1965), *Disc. Faraday Soc.*, **39**, 239.
[490] Bockris, J. O'M. (1965), *Disc. Faraday Soc.*, **39**, 261.
[491] Bockris, J. O'M. and Matthews, D. B. (1966), *J. Chem. Phys.*, **44**, 298.
[492] Bockris, J. O'M. and Matthews, D. B. (1966), *Electrochim. Acta*, **11**, 143.
[493] Dogonadze, R. R., Kuznetsov, A. M., and Levich, V. G. (1968), *Electrochim. Acta*, **13**, 1025.
[494] Dogonadze, R. R. (1971), in *Reactions of Molecules at Electrodes*, ed. N. S. Hush (Wiley, London).
[495] Volkenstein, M. V., Dogonadze, R. R., and Madumarov, A. K. (1971), *Dokl. Akad. Nauk S.S.S.R.*, **199**, 124.
[496] Dogonadze, R. R., Ulstrup, J., and Kharkats, Yu. I. (1972), *Dokl. Akad. Nauk S.S.S.R.*, **207**, 640.
[497] Dogonadze, R. R., Ulstrup, J., and Kharkats, Yu. I. (1974), *J. Chem. Soc. Faraday II*, **70**, 64.

[498] Kharkats, Yu. I. and Ulstrup, J. (1975), *J. Electroanal. Chem.*, **65**, 555.
[499] Krishtalik, L. I. and Tsionsky, V. M. (1971), *J. Electroanal. Chem. Interfacial Electrochem.*, **31**, 363.
[500] Appleby, A. J., Bockris, J. O'M., Sen, R. K., and Conway, B. E. (1972), *MTP Internat. Rev. Sci.*, Series 1, Vol. 6 (Butterworths, London) p. 1.
[501] Schmidt, P. P. and Ulstrup, J. (1973), *Nature (Phys. Sci.)*, **245**, 126.
[502] Appleby, A. J. (1974), *J. Electroanal. Chem. Interfacial Electrochem.*, **51**, 1.
[503] Bockris, J. O'M. and Sen, R. K. (1975), *Mol. Phys.*, **29**, 357.
[504] Narayanmurti, V. and Pohl, R. O. (1970), *Rev. Mod. Phys.*, **42**, 201.
[505] Barker, A. S. and Sievers, A. J. (1975), *Rev. Mod. Phys.*, **47**, 2nd supplement, p. 118.
[506] Anderson, P. W., Halperin, B. and Varma, C. (1972), *Phil. Mag.*, **25**, 1.
[507] Williams, J., Shohamy, E., Reich, S., and Eisenberg, A. (1975), *Phys. Rev. Lett.*, **35**, 951.
[508] Mavrin, B. N., Sterin, Kh. E., and Mischenko, A. V. (1973), *Fiz. Tverd. Tela*, **15**, 3702.
[509] Plesser, T. and Stiller, H. (1969), *Solid State Comm.*, **7**, 323.
[510] Villain, J. and Aubry, S. (1969), *Phys. Status Solidi*, **33**, 337.
[511] Birnbaum, H. K. and Wert, C. A. (1972), *Ber. Bunsenges. Phys. Chem.*, **76**, 802.
[512] Holleck, G. and Wicke, E. (1967), *Z. Phys. Chem. (Leipzig)*, **56**, 155.
[513] Ebisuzaki, Y., Kass, W. J., and O'Keeffe, M. (1967), *J. Chem. Phys.*, **46**, 1373.
[514] Stoneham, A. M. (1972), *Ber. Bunsenges. Phys. Chem.*, **76**, 816.
[515] Löwdin, P. O. (1963), *Rev. Mod. Phys.*, **35**, 724.
[516] Löwdin, P. O. (1965), *Adv. Quantum Chem.*, **2**, 213.
[517] Löwdin, P. O. (1965), *Mutat. Res.*, **2**, 218.
[518] Rein, R. and Harris, F. E. (1964), *J. Chem. Phys.*, **41**, 3393.
[519] Rein, R. and Harris, F. E. (1965), *J. Chem. Phys.*, **42**, 2177.
[520] Rein, R. and Harris, F. E. (1965), *J. Chem. Phys.*, **43**, 4415.
[521] Rein, R. and Ladik, J. (1964), *J. Chem. Phys.*, **40**, 2466.
[522] Ladik, J. (1964), *J. Theor. Biol.*, **6**, 201.
[523] Rai, D. K. and Ladik, J. (1968), *J. Molec. Spectrosc.*, **27**, 79.
[524] Ladik, J. (1973), *Adv. Quantum Chem.*, **7**, 397.
[525] Lunell, S. and Sperber, G. (1967), *J. Chem. Phys.*, **46**, 2119.
[526] Clementi, E., Mehl, J., and von Niessen, W. (1971), *J. Chem. Phys.*, **54**, 508.
[527] Pollard, E. and Lemke, M. (1965), *Mutat. Res.*, **2**, 213.
[528] Thomas, J. M., Evans, J. R. N., Lewis, T. J., and Secker, P. (1969), *Nature*, **222**, 375.
[529] Ionov, S. P. and Ionova, G. V. (1972), *Zh. fiz. Khim.*, **46**, 845; *Dokl. Akad. Nauk S.S.S.R.*, **202**, 960.
[530] Ionov, S. P. and Ionova, G. V. (1973), *Biofizika*, **18**, 205.
[531] Ionov, S. P., Lyubimov, V. S., and Ionova, G. V. (1974), *Zh. fiz. Khim.*, **48**, 1640.
[532] Goldanskii, V. I. (1977), *Nature*, **269**, 583.
[533] Greenberg, J. M. (1973), in *Molecules in the Galactic Environment* ed. M. A. Gordon and L. E. Snyder (Wiley, New York) p. 93.
[534] Greenberg, J. M. (1976), *Nederl. Tijdsch. f. Natuurkunde*, **42**, 117.
[535] Williams, D. A. (1971), *Astrophys. Lett.*, **10**, 71.

[536] Miller, T. J. and Williams, D. A. (1975), *Mon. Not. Roy. Astr. Soc.*, **173**, 527.
[537] Herbst, E. and Klemperer, W. (1976), *Phys. Today*, **29**, 32.
[538] Goldanskii, V. I. (1977), *Nature*, **268**, 612.
[539] Wickramasinghe, N. C. (1974), *Nature*, **252**, 452.
[540] Wickramasinghe, N. C. (1975), *Mon. Not. Roy. Astr. Soc.*, **170**, 11.
[541] Mendis, D. A. and Wickramasinghe, N. C. (1975), *Astrophys. Space Sci.*, **33**, 113.
[542] Hoyle, F. and Wickramasinghe, N. C. (1977), *Nature*, **268**, 610.
[543] Marcus, R. A. (1974), in *Techniques of Chemistry*, Vol. VI, Part 1 (Wiley – Interscience, New York) p. 13.
[544] Kramers, H. A. (1940). *Physica*, **7**, 284.
[545] Zwolinski, B. and Eyring, H. (1947), *J. Amer. Chem. Soc.*, **69**, 2702.
[546] Montroll, E. W. and Shuler, K. E. (1958), *Adv. Chem. Phys.*, **1**, 361.
[547] Present, R. D. (1959), *J. Chem. Phys.*, **31**, 747.
[548] Mahan, B. H. (1960), *J. Chem. Phys.*, **32**, 362.
[549] Noyes, R. M. (1961), *Progr. React. Kinet.*, **1**, 129.
[550] Marcus, R. A. (1966), *J. Chem. Phys.*, **45**, 2138.
[551] Morokuma, K. and Karplus, M. (1971), *J. Chem. Phys.*, 55, 63.
[552] Belloc, H. (1907), *Cautionary Tales for Children* (Duckworth, London).
[553] Marcus, R. A. (1964), *J. Chem. Phys.*, **41**, 2614, 2624.
[554] Marcus, R. A. (1965), *J. Chem. Phys.*, **43**, 1598.
[555] George, T. F. and Miller, W. H. (1972), *J. Chem. Phys.*, **57**, 2458.
[556] George, T. F. (1975), *J. Chem. Phys.*, **62**, 750.
[557] Christov, S. G. and Georgiev, Z. L. (1978), *J. Phys. Chem.*, **75**,1748.
[558] Bell, R. P. (1973), *The Proton in Chemistry*, 2nd edn. (Chapman and Hall, London) Chapters 4 and 5.
[559] Cohen, A. O. and Marcus, R. A. (1968), *J. Phys. Chem.*, **72**, 4249.
[560] Albery, W. J., Campbell-Crawford, A. N., and Curran, J. S. (1972), *J. Chem. Soc. Perkin II*, 2206.
[561] Bannister, C. E., Margerum, D. W., Raychera, J. M. T., and Wong, L. F. (1975), *Chem. Soc. Faraday Symp.*, **10**, 78.
[562] Hubbard, C. D., Wilson, C. J., and Caldin, E. F. (1976), *J. Amer. Chem. Soc.*, **98**, 1870.
[563] Dogonadze, R. R., Kuznetsov, A. M., and Levich, V. G. (1967), *Elektrokhimiya*, **3**, 739.
[564] Dogonadze, R. R., Kuznetsov, A. M., Levich, V. G., and Kharkats, Yu. I. (1970), *Electrochim. Acta*, **15**, 353.
[565] German, E. D., Dogonadze, R. R., Kuznetsov, A. M., Levich, V. G., and Kharkats, Yu. I. (1971), *J. Res. Inst. Catal. Hokkaido Univ.*, **19**, 99, 115.
[566] Dogonadze, R. R. and Kuznetsov, A. M. (1971), *Elektrokhimiya*, **6**, 763.
[567] Vorotintsev, M. A., Dogonadze, R. R., and Kuznetsov, A. M. (1973), *Dokl. Akad. Nauk S.S.S.R.*, **209**, 1135.
[568] Vorotintsev, M. A., Dogonadze, R. R., and Kuznetsov, A. M. (1974), *Elektrokhimiya*, **10**, 687, 867.
[569] Dogonadze, R. R. and Kuznetsov, A. M. (1975), *J. Res. Inst. Catal. Hokkaido Univ.*, **22**, 93.
[570] German, E. D. and Kharkats, Yu. I. (1973), *Izv. Akad. Nauk. Ser. Khim.*, 1031.

[571] German, E. D. and Dogonadze, R. R. (1977), Appendix to *Proton v Khimii* by R. P. Bell (Mir, Moscow) pp. 350–376.
[572] Smith, D. W. G. and Powles, J.G. (1966), *Mol. Phys.*, **10**, 451.
[573] Hertz, H. G. (1970), *Angew. Chem. Int. Ed.*, **9**, 124.
[574] Eisenberg, D. and Kauzmann, W. (1969), *The Structure and Properties of Water* (Clarendon Press, Oxford) p. 214.
[575] Pimentel, G. C. and McClellan, A. L. (1971), *Ann. Rev. Phys. Chem.*, **22**, 347.
[576] Hirschfelder, J. O., Curtis, C. F., and Bird, R. B. (1954), *Molecular Theory of Gases and Liquids* (Wiley, New York).
[577] Bell, R. P. (1980), to be published.

# Author Index

(Reference numbers are given first, in square brackets, followed by page numbers)

Abraham, R.J., [444, 445, 449]; 158
Ahlberg, P., [306–310]; 130
Alberding, N., [373, 379]; 142, 143
Albery, W.J., [172, 560]; 93, 177
Allen, J.D., [316, 353]; 130, 136
Alpert, N.L., [416]; 155
Anbar, M., [165]; 92, 131
Anderson, P.W., [506]; 165
Andrew, E.R., [418]; 155
Antonovskii, V.L., [333]; 131
Apaydin, F., [420]; 155
Appleby, A.J., [500, 502]; 164
Arthur, N.L., [304]; 127
Ashmore, P.G., [291]; 125
Aubry, S., [510]; 165
Austin, R.H., [370–373, 379]; 142

Baer, M., [292–294]; 125, 168
Band, Y.B., [203]; 99
Banger, J., [190]; 96
Bannister, C.E., [561]; 177
Barclay, L.R.C., [231–233]; 110, 140, 173
Barkalov, I.M., [362–369]; 141, 166
Barker, A.S., [505]; 165
Barnes, D.J., 159]; 92, 96, 129
Barrow, G.M., [400]; 154
Bar-Yaakov, Y., [287]; 124
Bawn, C.E.H., [41]; 11, 164
Beddard, G.S., [394]; 150
Beeson, K.W., [370–373]; 142
Bell, R.P., [40, 50, 51, 68, 81, 93, 102–105, 108, 109, 111, 117, 147, 152, 155–159, 173, 180, 187, 188, 201, 218, 352, 447, 558, 577]; 11, 32, 33, 41, 60, 63, 71, 73, 74, 77, 82, 91–93, 95–97, 102, 106, 118, 127, 128, 130, 136, 138, 158, 176, 184, 187, 193
Belloc, H., [552]; 169
Bengtsson, S., [309]; 130

Berezin, I.V., [333, 335]; 131
Bergman, N., [209, 442]; 103, 139, 157, 178
Bernstein, H.J., [434]; 156
Berry, R.S., [67, 462]; 49, 162
Bersohn, R., [418, 425]; 155
Bethell, D., [329]; 131
Bigeleisen, J., [121, 132, 150, 170, 284]; 77, 84, 91, 93, 124
Bird, R.A., [166]; 92
Bird, R.B., [576]; 184
Birnbaum, H.K., [511], 165
Blackwell, L.F., [191, 194, 330]; 96, 97, 130, 131
Blanch, J.H., [347]; 133, 179, 181
Boato, G., [257]; 117
Bockris, J. O'M., [487–492, 500, 503]; 164
Bolshakov, A.I., [378]; 142
Bonin, M.A., [225]; 107
Bordwell, F.G., [157]; 92, 96
Born, M., [3, 35, 37]; 6, 11
Bourgin, D.G., [33]; 11
Bowden, K., [192]; 96
Bowen, T., [381]; 144
Bowman, D.F., [341]; 131
Bowman, J.M., [298]; 126
Bowman, N.S., [116]; 77
Boyle, W.J., [157]; 92, 96
Brewer, J.H., [380, 381]; 143, 144
Brickmann, J., [75, 76, 410–413, 451]; 50, 154, 160
Brillouin, L., [55, 56]; 34
Bromberg, A., [242–245]; 113, 114
Bron, J., [135]; 87
Brown, N.M.D., [25]; 9
Bruice, T.C., [160]; 92, 96
Brunton, G., [232, 233]; 110, 140, 173
Buckley, P.D., [330]; 131
Bunnett, J.F., [184]; 96
Burr, J.C., [443]; 157

Burstein, E., [18]; 9
Busch, J.H., [79]; 50, 154
Butler, M.M., [334]; 131
Butyagin, P.Yu., [241]; 113

Caldin, E.F., [195, 196, 321–326,
  346–349, 355–358, 439, 562]; 103,
  130, 133, 136, 138, 139, 156, 179,
  180
Caldwell, J.D., [320]; 130
Campbell-Crawford, A.N., [560]; 177
Campion, A., [222]; 109, 174
Careri, G., [257]; 117
Carolan, J.L., [419]; 155
Carter, R.E., [442]; 157
Castro, J.C., [7]; 7
Chan, S.S., [370, 373, 379]; 142, 143
Chanmugan, J., [291]; 125
Chapman, S., [90]; 62, 123
Chen, H.L., [201]; 97, 177
Chiang, Y., [130, 131]; 82
Child, M.S., [277]; 122, 168
Chiltz, G., [285]; 124
Chin, K.W., [11]; 8
Cholod, M.S., [208]; 102
Christoph, A.C., [13]; 8
Christov, S.G., [66, 99–101, 468–475,
  557]; 41, 69, 70, 141, 164, 171
Cimino, A., [257]; 117
Clark, T.C., [301]; 126
Cleeton, C.E., [399]; 151
Clementi, E., [526]; 165
Clough, S., [420, 423]; 155
Cockerill, A.F., [186, 329]; 96, 131, 179
Cohen, A.O., [559]; 176
Coleman, R.V., [23, 24]; 9
Collins, C.J., [116]; 77
Coltrin, M.E., [272]; 121, 171
Condon, E.U., [27, 28]; 9
Connor, J.N.L., [274]; 122, 168
Connor, T.M., [437]; 156
Conway, B.E., [480–486, 500]; 164
Cook, D., [161, 185]; 92, 96
Coon, D.D., [5]; 7
Costain, C.C., [402]; 154
Coté, G.L., [126]; 82
Cox, B.G., [187, 188, 193]; 96, 97, 179
Cram, D.J., [316]; 130, 140
Cremer, E., [39]; 11
Critchlow, J.E., [447]; 158
Crooks, J.E., [155]; 96
Crowe, K.M., [380, 381]; 143, 144
Curran, J.S., [560]; 177
Curtis, C.F., [576]; 184

Davidson, D.W., [421]; 155
Davidson, R.B., [432]; 156
De Broglie, L., [1]; 3
Debrunner, P.G., [370]; 142
De Haas, N., [260]; 117
De la Vega, J.R., [78, 79]; 50, 154
Delise, D.A., [381]; 144
Dennison, D.M., [71, 422]; 44, 151, 155
De Ribaupierre, Y., [215]; 104, 165
Dixon, J.E., [160]; 92, 96
Dogonadze, R.R., [493–497, 563–571];
  62, 164, 181
Doll, J.D., [279]; 123, 168
Donchi, K.F., [304]; 127
Douglas, A.E., [389]; 150
Dove, J.E., [301]; 126
Downing, R., [370]; 142
Downs, J.J., [466, 467]; 162
Drakenberg, T., [442]; 157
Dubinskaya, A.M., [241]; 113
Duke, C.B., [17, 46]; 9, 31
Dunham, J.L., [64]; 36
Dunn, M.B., [427]; 156

Earls, D.W., [189]; 96
Eaton, G.R., [145]; 90, 140, 158
Eaton, S.S., [145]; 90, 140, 158
Ebisuzaki, Y., [513]; 165
Eckart, C., [45]; 27
Eckling, R., [285]; 124
Eisenberg, A., [507]; 165
Eisenberg, D., [574]; 182
Eisenstein, L., [370–373, 379]; 142, 143
Emsley, J.W., [440]; 156
England, R.J., [226]; 107
Enke, C.G., [485]; 164
Evans, J.R.N., [528]; 166
Excoffon, P., [403]; 154
Eyring, H., [545]; 168

Farkas, A., [245, 256]; 117
Farkas, L., [256]; 117
Feeney, J., [440]; 156
Fendley, J.A., [218]; 106, 130
Feynman, R.P., [8]; 7
Fischer, E., [242, 243]; 113
Fischer, H.F., [315]; 130, 140
Fitos, I., [168]; 92, 131, 133
Flanigan, M.C., [78]; 50, 154
Fleischer, E.B., [454]; 160
Fleming, D.G., [380, 381]; 143, 144
Fleming, G.R., [394]; 150
Formosinho, S.J., [395]; 150
Fowler, R.H., [15]; 8, 32
Franck, J., [35]; 11

# Author index

Frank-Kamenetskii, M.D., [366, 367]; 141, 166
Frauenfelder, H., [370–373, 379]; 142, 143
Freed, J.H., [431]; 156
Freed, K.F., [203]; 99
French, W.G., [237]; 113
Friedlander, G., [30]; 10
Fröman, N., [63]; 35
Fröman, P.O., [63]; 35
Funderburk, L., [317, 318]; 130, 140
Furhop, J.H., [450]; 158

Galitzkii, V.M., [48]; 32
Gamow, G., [26]; 10
Garg, S.K., [421]; 155
Garner, D.M., [381]; 144
Garrett, B.C., [90]; 62, 123
Geib, K.H., [255]; 117
Geiger, H., [29]; 10
George, T.F., [279, 555, 556]; 123, 168, 170
Georgiev, Z.L., [557]; 171
German, E.D., [210, 565, 570, 571]; 103, 181
Gibson, A., [193]; 97, 179
Gijzeman, O.L.J., [394]; 150
Gillan, T., [341]; 131
Gilliom, R.D., [336]; 131
Ginn, S.W.G., [404]; 154
Goldanskii, V.I., [360–369, 376–378, 532, 538]; 141, 142, 166
Goldfinger, P., [285]; 124
Goldman, I.I., [48]; 32
Goldman, I.M., [345]; 131
Goodall, D.M., [158]; 92, 96, 130
Goos, F., [10]; 8
Grainger, S., [111]; 74
Gray, J.A., [233]; 110, 140, 173
Greenberg, J.M., [533, 534]; 166
Griller, D., [231–233]; 110, 140, 173
Grinstein, R.H., [176, 211]; 94, 103, 131
Guil, J.M., [216]; 104, 165
Gunsalus, I.C., [371, 372, 379]; 142, 143
Gurney, R.W., [27, 28]; 9
Gust, D., [446]; 158
Gutowsky, H.S., [425, 461]; 155, 162
Gygax, F.N., [380]; 143

Haas, C., [407]; 154
Hadzi, D., [406]; 154
Halavee, U., [294]; 125, 168
Halperin, B., [506]; 165

Hamor, M.J., [456]; 160
Hamor, T.A., [456]; 160
Hamrick, P.J., [246]; 114
Hänchen, M., [10]; 8
Hanna, S.B., [163]; 96
Hansma, P., [24]; 9
Harbron, E., [356]; 138
Harmony, M.D., [73, 80]; 47, 50
Harris, F.E., [518–520]; 165
Harteck, P., [255]; 117
Hassid, A.I., [200]; 177
Haven, Y., [246]; 114
Hawkes, G.E., [444, 445, 449]; 158
Hawthorne, M.F., [175]; 94
Hayward, D.O., [216]; 104, 165
Hecht, K.T., [422]; 155
Heicklen, J., [95]; 63
Heidberg, J., [441]; 156
Heisenberg, W.K., [2]; 4
Hennig, J., [448]; 140, 158, 175
Herbst, E., [537]; 166
Hertz, H.G., [573]; 182
Herzberg, G., [382, 383, 390, 396]; 146, 150, 151
Hill, D.L., [44]; 23
Hine, J., [208]; 102
Hirota, K., [401]; 154
Hirschfelder, J.O., [13, 576]; 8, 184
Hoard, J.L., [456]; 160
Hobson, T., [423]; 155
Holleck, G., [512]; 165
Hollinden, G.A., [300]; 126
Holmes, R.R., [463]; 162
Hornig, D.F., [77, 407]; 50, 154
Hoyle, F., [542]; 166
Hubbard, C.D., [562]; 180
Hudson, R.L., [223]; 107, 111, 174
Huebner, H., [312]; 135
Hulett, J.R., [218]; 106, 130
Hund, F., [31]; 11
Hunt, G.R., [393]; 150
Hutchinson, R.E.J., [161, 185]; 92, 96
Huybrechts, G., [285]; 124

Ikeda, R., [427, 428]; 156
Ingold, K.U. [231–233, 337, 341]; 110, 131, 140, 173
Ionov, S.P., [529–531]; 166
Ionova, G.V., [529–531]; 166
Isaacs, N.S., [212]; 104
Iwasaki, M., [235]; 112, 175

Jackson, R.A., [357]; 138
Jacox, M.E., [228]; 108
Jaffe, A., [190]; 96

Jaklevic, R.C., [21, 22]; 9
Jarczewski, A., [322, 351]; 130, 136
Javaid, K., [212]; 104
Jeffreys, B., [65]; 38
Jeffreys, H., [59]; 35
Jermini, C., [163]; 92, 96
Johns, J.W.C., [388]; 150
Johnson, C.S., [424]; 155
Johnson, R.E., [466, 467]; 162
Johnson, R.F., [380]; 143
Johnston, H.S., [43, 82–85, 89, 94, 95, 129, 248, 285, 290, 299]; 13, 54–57, 62, 63, 82, 117, 120, 123–127, 129, 170
Jolley, K.W., [330]; 131
Jones, J.R., [164, 189, 207, 313, 314]; 92, 96, 102, 130

Kampschultesche, I., [408]; 154
Kanzig, W., [430]; 156
Kaplan, A.M., [362–365]; 141, 166
Kardos, J., [168]; 92, 131, 133
Karplus, M., [87, 265, 275, 551]; 57, 98, 118, 122, 168
Kasparian, M., [358]; 138
Kass, W.J., [513]; 165
Katz, A.M., [181]; 96
Kaufman, J.C., [208]; 102
Kauzmann, W., [574]; 182
Keefe, J.R., [162]; 92, 96, 130, 136
Kelm, H., [213]; 104
Kemble, E.C., [60, 61, 70]; 37, 41, 43
Kennedy, J.W., [30]; 10
Kenney, C.N., [398]; 151
Khairutdinov, R.F., [376]; 142
Kharkats, Yu. I., [496–498, 564, 565, 570]; 164, 181
Kim, J.H., [350]; 136
Kiryukhin, D.P., [362–365, 368, 369]; 141, 166
Klein, F.S., [243, 284, 286–288]; 113, 124, 125
Klemperer, W., [537]; 166
Kneipp, K.G., [167]; 92
Koeppl, G.W., [271, 297]; 120, 126
Kogan, V.I., [48]; 32
König, J., [339]; 131
Kooter, J.A., [459]; 161
Kouba, J., [151]; 91, 94, 138
Kovács, I., [168]; 92, 131, 133
Kramers, H.A., [58, 544]; 35, 168
Kreevoy, M.M., [146, 198–200]; 90, 97, 177
Kresge, A.J., [106, 107, 130, 131, 201]; 74, 82, 97, 177

Krishtalik, L.I., [499]; 164, 181
Krivchenkov, U.D., [48]; 32
Krumbiegel, P., [311]; 135
Kuppermann, A., [86, 273, 298]; 57, 98, 121, 126, 168, 172
Kurylo, M.J., [300]; 126
Kurz, J.L., [359]; 139, 180
Kurz, L.C., [359]; 139, 180
Kuzina, S.I., [377]; 142
Kuznetsov, A.M., [92, 493, 563–569]; 62, 164, 181

Ladik, J., [521–524]; 165
Laidler, K.J., [88, 148]; 57, 91
Lambe, J., [21, 22]; 9
Landau, L.D., [42]; 12
Langer, R.M., [34]; 11
Larson, G.F., [336]; 131
Lebedev, Ya. S., [234, 378]; 112, 142
Lee, D.G., [344]; 131
Leffek, K.T., [322, 350, 351]; 130, 136
Lemke, M., [527]; 166
Le Roy, D.J., [49, 52, 53, 258, 259, 261–263, 267, 268]; 32, 34, 60, 62, 117, 119
Le Roy, R.J., [52–54, 220]; 34, 60, 108, 119, 174
Lester, W.A., [270]; 120, 172
Levich, V.G., [493, 563–565]; 164, 181
Lewis, E.S., [174–176, 205, 211, 316–320, 334, 353]; 94, 101, 103, 130, 131, 136, 140
Liang, T.-M., [200]; 97, 177
Lifschitz, G., [331]; 131
Lifshitz, E.M., [42]; 12
Limbach, H., [448]; 140, 158, 175
Lin, A.-C., [190]; 96
Lin, C.C., [414]; 154
Lindemann, R., [409]; 154
Liu, B., [269]; 120, 172
Loewenschuss, H., [163]; 92
Loewenstein, A., [436, 437]; 156
London, F., [247]; 117, 125
Long, F.A., [154]; 92, 96
Longuet-Higgins, H.C., [123]; 82
Lotsch, H.K.V., [12]; 8
Löwdin, P.O., [74, 515–517]; 48, 96, 165
Lukowits, I., [168]; 92, 131, 133
Lukovnikov, A.F., [377]; 142
Lundqvist, S., [18]; 9
Lunell, S., [525]; 165
Lutz, M., [457]; 160
Lyubimov, V.S., [531]; 166

# Author index

McCall, A.D., [461]; 162
McClellan, A.L., [575]; 182
McCoy, E.F., [392, 393]; 150
McDonald, W.J., [6]; 7
McDonell, J.A., [304]; 127
McDowell, C.A., [427, 428]; 156
McGibbon, A.K.H., [330]; 131
Mackinnon, D.J., [486]; 164
Macleod, J.K., [161, 185]; 92, 96
McLean, S., [340]; 131
McNesby, J.R., [303]; 127
Madumarov, A.K., [495]; 164, 181
Mahan, B.H., [551]; 168
Mains, G.J., [264]; 117
Manchester, F.D., [215]; 104, 165
Manning, M.F., [72]; 47, 153
Marcus, R.A., [91, 197, 272, 543, 550, 553, 554, 559]; 62, 97, 121, 167–169, 171, 176, 177
Maréchal, Y., [403]; 154
Margerum, D.W., [561]; 177
Marks, R.E., [314]; 130
Marshall, V.P., [371]; 142
Martin, B., [328]; 130
Mason, S., [452]; 160
Mateo, S.,[195, 196, 323]; 97, 103, 130, 136, 139, 179, 180
Matthews, D.B., [489, 491, 492]; 164
Mavrin, B.N., [508]; 165
Mayer, I., [217]; 105
Mayer, M.G., [121]; 77, 84
Mehl, J., [526]; 165
Meiboom, S., [435]; 156
Merritt, J.A., [443]; 157
Melander, L.,[112, 209]; 77, 102, 139, 178
Mendis, D.A., [541]; 166
Meyers, L., [285]; 124
Meyerstein, D., [165]; 92, 131
Mikhailov, A.I., [376–378]; 142
Miller, J.M., [30]; 10
Miller, T.J., [536]; 166
Miller, W.H., [90, 278–282]; 62, 123, 168, 170–172
Milligan, D.E., [228]; 108
Millington, J.P., [147]; 91
Minn, F.L., [265]; 119
Mischenko, A.V., [508]; 165
Mitchell, D.N., [261]; 117
Mitchell, R.W., [443]; 157
Miyagawa, I., [432]; 156
Mocek, M.M., [343]; 131
Molinari, E., [257]; 117
Montroll, E.W., [546]; 168
More O'Ferrall, R.A., [151, 177]; 91, 94, 138

Morokuma, K., [551]; 168
Mortensen, E.M., [276]; 122, 168
Mottley, C., [424]; 155
Muetterties, E.L., [460, 464, 465]; 162
Mulliken, R.S., [384]; 146
Munderloh, N.H., [162]; 92
Murrell, J.N., [148]; 91
Muszkat, K.A., [242, 243, 245]; 113, 114

Nakai, Y., [401]; 154
Narayanmurti, V., [504]; 165
Neiss, M.A., [238, 239]; 113
Newton, Sir Isaac, [4]; 7
Niki, H., [264]; 117
Noe, L.J., [229]; 108
Nordheim, L., [15, 16]; 8, 32
Nordlund, T.M., [370, 373, 379]; 142
Noyes, R.M., [549]; 168
Nugent, S.M., [423]; 155
Nunome, K., [235]; 112, 175
Nuttall, J.M., [29]; 10

Ogden, G., [41]; 11, 164
Oh, S., [198]; 97
O'Keeffe, M., [513]; 165
Olsson, E., [385]; 149
Onwood, D.P., [352]; 136
Oppenheimer, J.R., [14, 32]; 8, 11

Pace, E.L., [229]; 108
Pacey, P.D., [302]; 126
Palke, W.E., [13]; 8
Palmer, D.A., [213]; 104
Parboo, D.M.,[325, 326]; 130, 136, 179
Parker, A.J., [161, 185]; 92, 96
Parr, C., [290]; 125
Pearson, R.G., [125]; 82
Perry, J.M., [211]; 103, 131
Persky, A., [286–288, 292, 294–296]; 124–126, 168
Petch, H.E., [426]; 155
Phillips, W.D., [464, 465]; 162
Pifer, A.E., [381]; 144
Pimentel, G.C., [575]; 182
Pink, J.M., [147]; 91
Pintar, M.M., [426]; 155
Pitzer, K.S., [276]; 122, 168
Plesser, T., [509]; 165
Pohl, H.A., [9]; 7
Pohl, R.O., [504]; 165
Polanyi, M., [39]; 11
Pollard, E., [527]; 165
Pople, J.A., [434]; 156
Porter, G., [394]; 150

Porter, R.N., [87, 275]; 57, 98, 122, 168
Pospišil, J., [168]; 92, 131, 133
Powles, J.G., [572]; 182
Present, R.D., [547]; 168
Primack, W., [374, 375]; 142
Pruszynski, P., [351]; 136
Pryor, W.A., [167]; 92

Quickert, K.A., [49, 52, 53, 263, 267, 268]; 32, 34, 62, 117, 119
Quinn, J.J., [11]; 8

Rai, D.K., [523]; 165
Rannala, E., [212]; 104
Rapp, D., [89]; 62, 120, 123, 125, 127, 170
Raychera, J.M.T., [561]; 177
Redlich, O., [133]; 85
Reich, S., [507]; 165
Rein, R., [518–520]; 165
Reitz, O., [153]; 92, 96
Reuwer, J.F., [149]; 91
Ridley, B.A., [259, 263]; 117
Ripmeester, J.A., [421]; 155
Roberts, J.D., [433, 446]; 156, 158
Roberts, R.J., [337]; 131
Robinson, G.W., [391]; 150
Robinson, J.K., [205, 319]; 101, 130
Rock, P.A., [118]; 77
Roginsky, S., [36]; 11
Rogne, O., [346–348]; 133, 179, 181
Rollefson, G.K., [283]; 124
Rosenkewitsch, L., [36]; 11
Ross, I.G., [392, 393]; 150
Roth, W.R., [339]; 131
Rothschild, W.G., [405]; 154
Rummel, S., [312]; 135
Rumney, T.C., [189]; 96
Russell, K.E., [166]; 92
Rutherford, R.J.D., [340]; 131

Sachs, W.H., [105]; 73, 82, 95
Sagatys, D.S., [201]; 97, 177
Salomon, M., [481–485]; 164
Sato, S., [289]; 125, 171
Saunders, W.H., [115, 181–183, 190]; 77, 96
Schaad, L.J., [149]; 91
Schatz, G.C., [273, 298]; 121, 126, 168, 172
Schawlow, A.L., [397]; 151
Schenk, A., [380]; 143
Schlichter, C.P., [461]; 162
Schmidt, P.P., [501]; 164
Schneider, M.G., [141, 143]; 90

Schneider, W.G., [434]; 156
Schulz, W.R., [258, 259, 262]; 117
Scott, T.A., [419]; 155
Sen, R.K., [500, 503]; 164
Shapiro, J.S., [128]; 82
Sharma, R.D., [87]; 57, 98, 122
Sharp, A.R., [426]; 155
Sharp, T.E., [129]; 82, 126, 129
Shavitt, I., [265, 266]; 118
Shields, H., [246]; 114
Shin, H., [96]; 63
Shiner, V.J., [327, 328]; 130
Shiotani, M., [223]; 107, 111, 174
Shishkina, L.N., [325]; 130
Shohamy, E., [507]; 165
Shuler, K.E., [546]; 168
Sicking, G., [214]; 104, 165
Sievers, A.J., [505]; 165
Silvers, S.J., [453]; 160
Simonsen, M.J., [23, 24]; 9
Simonyi, M., [168, 217, 338]; 92, 105, 131, 133
Simpson, C.J.S.M., [124]; 82
Smith, D., [429]; 156
Smith, D.W.G., [572]; 182
Smith, I.W.M., [305]; 127
Smith, K.M., [444, 445, 449]; 158
Smith, M.L., [327]; 130
Soddy, F., [119]; 77
Sokolski, E.A., [144]; 90, 158
Solymar, L., [19]; 9
Soma, M.J.J., [459]; 161
Somorjai, R.L., [77]; 50
Sperber, G., [525]; 165
Sprague, E.D., [54, 219, 220, 224, 239]; 34, 107–109, 113, 174
Srinivasan, R., [415]; 155
Srinivasan, S., [487–489]; 164
Srivastava, G.P., [402]; 154
Steacie, E.W.R., [230]; 109
Stein, G., [234]; 112
Sterin, Kh.E., [508]; 165
Stern, M.J., [97, 136–143, 204, 206, 288]; 67, 90, 100, 101, 125
Stevens, R.M., [265]; 119
Stewart, R., [178, 342–344]; 94, 131
Stiller, H., [509]; 165
Stivers, E.C., [149]; 91
Stoneham, A.M., [514]; 165
Storm, B., [144]; 90, 158
Strehlow, H., [438]; 156
Stwalley, W.C., [386, 387]; 149
Subba Rao, S.C., [314]; 130
Sugden, T.M., [398]; 151
Sutcliffe, L.H., [440]; 156

# Author index

Swain, C.G., [149]; 91
Swalen, J.D., [414]; 154
Symons, M.C.R., [174, 226]; 94, 107

Taylor, N., [216]; 104, 105
Teklu, Y., [144]; 90, 158
Ter Haar, D., [47]; 31
Thibblin, A., [308–310]; 130
Thomas, J.M., [528]; 166
Thomas, J.T., [416]; 155
Thompson, H.W., [125]; 82
Tilak, B.V., [486]; 164
Timmons, R.S., [300]; 126
Titchmarsh, E.C., [62]; 35
Tolman, R.C., [97]; 64
Tomalin, G., [321]; 130, 136
Tomita, K., [417]; 155
Torrey, H.C., [416]; 155
Toryiama, K., [235]; 112, 175
Townes, C.H., [397]; 151
Tranter, R.L., [105, 156]; 73, 82, 92, 95, 96, 129
Trotman-Dickenson, A.F., [230]; 109
Truhlar, D.G., [86, 251, 252]; 57, 98, 117
Tsionsky, V.M., [499]; 164, 181
Tsuikow-Roux, E., [299]; 126
Tsuji, K., [255]; 107
Tüdös, F., [338]; 131
Tulinsky, A., [453, 455]; 160

Uhlenbeck, G.E., [71]; 44, 151
Ulstrup, J., [486–498, 501]; 164, 181
Urey, H.C., [120]; 77, 84

Vacek, K., [332]; 131
Van der Linden, R., [342]; 131
Van der Waals, J.H., [458, 459]; 161
Van Dorp, W.G., [458, 459]; 161
Van Hook, W.A., [127, 250]; 82, 117
Varma, C., [506]; 165
Verbekel, G., [285]; 124
Villain, J., [510]; 165
Vogel, P.C., [140–142]; 90, 100
Volkenshtein, M.V., [495]; 164, 181
Volpi, G.G., [257]; 117
Von Laue, M., [69]; 43
Von Niessen, W., [526]; 165
Von Sonntag, C., [332]; 131
Walker, F.A., [325]; 130, 136
Wallick, E.T., [353]; 136
Walmsley, D.G., [25]; 9
Wang, J.T., [221]; 107, 174

Warshel, A., [244, 245]; 114
Watson, D., [154]; 92, 96
Watton, A., [426]; 155
Way, K.R., [386]; 149
Webb, L.E., [454]; 160
Webster, C.J., [340]; 131
Weiss, J.J., [110]; 74
Weisskopf, V., [37]; 11
Wentzel, G., [57]; 34
Wert, C.A., [511]; 165
Westenburg, A.A., [260]; 117
Westheimer, F.H., [169]; 92, 93
Weston, R.E., [98, 128, 204, 206, 249, 284]; 67, 82, 101, 117, 124
Wheeler, J.A., [44]; 23
Wicke, E., [512]; 165
Wickramasinghe, N.C., [539–542]; 166
Wigner, E., [38]; 11, 53, 61, 63, 64
Wijnen, M.H.J., [227]; 108
Wilkey, D.D., [240]; 113
Willard, J.E., [236–240]; 113
Willi, A.V., [171, 179, 354]; 93, 94, 136
Williams, D.A., [535, 536]; 166
Williams, F., [54, 219–223, 225]; 34, 107, 111, 174
Williams, J., [507]; 165
Williams, N.H., [399]; 151
Williams, R.C., [246]; 114
Wilson, C.J., [324–326, 562]; 130, 136, 179, 180
Wilson, H., [320]; 130
Wolf, E., [3]; 6
Wolfsberg, M., [134–139, 171, 284]; 87, 90, 93, 124
Wong, L.F., [561]; 177
Wong, S.M., [315]; 130, 140
Wood, J.L., [404]; 154
Woodhead, J.L., [191, 194]; 96, 97, 131
Wyatt, R.E., [251, 252]; 117

Yakinchenko, O.E., [234]; 112
Yates, A.C., [270]; 120, 172

Zamaraev, K.I., [376]; 142
Zellner, R., [305]; 127
Ziman, J.M., [20]; 9
Zimmermann, H., [75, 76, 410–412]; 48, 49, 154
Zollinger, H., [122, 163]; 82, 92, 96, 129
Zundel, G., [408, 409]; 154
Zwolinski, B., [545]; 168

# Subject

Acetic acid, 138, 154
Acetone, 102
Acetonitrile, 97, 107
1-(2-Acetoxy-3-propyl)-indene, 130
Action, 36
Airy functions, 31
'All-or-nothing' isotope effects, 109, 112
Alpha-particle emission from nuclei, 9
Ammonia inversion, 14, 44, 150
Ammonia, p.m.r. spectrum of, 155
Ammonium salts, p.m.r. spectra of, 156
Aromatic substitution, 81, 129
Arrhenius equation, 64
    deviations from, 66, 108, 138, 141, 159, 175
Arrhenius parameters, 64, 89, 99
Arsine, inversion of, 153
Aryl radicals, isomerization of, 110, 173
Autoxidation, 135
Aziridine inversion, 156

Barrier frequency (imaginary), 25, 53, 58, 115
Barriers, dimensions of, 137
    permeability of, 1
Bending vibrations, 82
Benzyl $t$-butyl ether, 135
Bessel functions, 31, 189
Bohr theory, 36
Bond orders, 94
Born–Oppenheimer approximation, 84, 178
Bound particles, tunnelling by, 14, 42
Bromocyclohexyl bromide, 96
Brönsted relation, 74, 92
BWK (Brillouin-Wentzel-Kramers) approximation, 34, 45, 123

Carbon acids, 95
Carbon monoxide, tunnelling by, 142
Chain reactions, 135
Characteristic temperature for tunnelling, 69, 141

Chloranil, 103
Chlorine atoms, reaction with hydrogen, 124
Chlorine trifluoride, n.m.r. spectrum of, 162
Complex variables, 24, 187
Computer experiments, 90
Concerted processes, 90
Confluent hypergeometric functions, 187
Crown ether, 140
Crystal violet, 103
Cumene, 135
Cut-off procedure, 90
Cytosine, 165

De Broglie wavelength, 3, 21, 98, 169
Diffusion of hydrogen in metals, 104, 165
4a-4b-Dihydrophenanthrene, 113
Dimethyl ether, p.m.r. spectrum of, 155
Dimethyglyoxime, 112, 175
Dimethyl sulphide, p.m.r. spectrum of, 155
Dimethyl sulphoxide, 96
Di-(4-nitrophenyl)-methane, 136
1, 1-Diphenylethane-2-benzene-sulphonate, 136
Double anharmonic barrier, 32, 40, 188
Double-minimum potentials, 43, 49

Eckart barrier, 27, 63, 108
Electrode processes, 163
Electron emission from metals, 8
Electron spin resonance spectra, 49, 107, 156
Electron tunnelling in solids, 9
Electrostatic models, 72, 95
Elimination reactions, 95, 128
Encounter complex 177
Energy surfaces, 118
2-Ethoxycarbonylcyclopentanone, 106
Ethyl vinyl ether, 82
Ethyleneimine inversion, 156

# Subject index

Fluorine atoms, reaction with hydrogen, 126
    tunnelling by, 162
Formaldehyde polymerization, 141
Formic acid, 83, 136, 154
Franck-Condon principle, 177
Free-particle tunnelling, 13
Frustrated total reflection, 7

Gaussian barriers, 34, 108, 109
Geiger-Nuttall relationship, 10
Goos-Hänchen effect, 8
Guanine, 165

Heavy-particle tunnelling, 140
Heisenberg uncertainty principle, 4, 50, 62, 84
Hemoglobin, 142
Hindered rotation, 154
Hydrofluoric acid, 83, 138
Hydrogen atom transfers,
    at low temperatures, 107
    in gas reactions, 114
Hydrogen, cathodic discharge of, 164, 181
    diffusion in metals, 104, 165
Hydrogen bonds, tunnelling in, 153, 165
Hydroxyurea, 114
Hypergeometric functions, 29, 32, 186, 194

Ice, 154
Imaginary frequency, 25, 86, 91, 115, 119, 137
Intermolecular forces, 183
Intramolecular processes, 14

Life, origin of, 166
Lorentzian barriers, 166

Mathieu functions, 41
Metals, diffusion of hydrogen in, 104, 165
    electron emission from, 8, 29
Methane, p.m.r. spectrum of, 155
Methanol, 109, 174
Methyl group rotation, 14, 154
Methyl isocyanide, 109
3-Methylpentane, 113
Microwave spectra, 151
Morse barrier, 32, 40, 188
Muonium, 143
Mutarotation, 128
Mutations, 165
Myoglobin, 143

Newton's rings, 7
Nitric oxide, tunnelling by, 143
4-Nitrophenylnitromethane, 97, 133, 139, 179
Nuclear emission of $\alpha$-particles, 9

Optical analogies of tunnel effect, 5
Ortho- and para-hydrogen, 117

Palladium, diffusion of hydrogen in, 104
Parabolic barrier, 21, 40, 60, 108, 193
Parabolic cylinder functions, 23
Particle density and flux, 16
Permeability of barriers, 1
    numerical calculation of, 34
Perturbation theory, 16
Phenylphosphinic acid, 154
Phosphine inversion, 153
Phosphorus pentafluoride, n.m.r. spectrum of, 162
Photodissociation, 99
Polychromatic kinetics, 142
Polyepoxyethane, 113
Polyisobutene, 113
Polyoxymethylene, 166
Potassium dihydrogen phosphate, 154, 165
Potassium hydrogen di-(4-nitrobenzoate), 154
Predissociation,
    by radiationless transition, 150
    by rotation, 146
    of diatomic hydrides, 146
Pre-exponential factors, 63
Pressure effects, 103
Product rules, 85
Propane, p.m.r. spectrum of, 155
Proteins, proton conduction in, 166
Protoheme, 143
Proton magnetic resonance spectra, 155

Racemization, 128, 140
Radioactivity, 9
Reaction complex, 177
Reaction co-ordinate, 12
    separability of, 169
Rectangular barriers, 17, 39
Reduced mass, 12, 56, 79, 104, 120, 170, 178

Semi-classical approximation, 34
Separability of reaction co-ordinate, 169
Solvent reorganization, 103, 176
Stationary states, 47

Statistical factors, 87
Steady state treatment, 15
Steric hindrance, 140, 184
Sulphur tetrafluoride, n.m.r. spectrum of, 162
Swain-Schaad relation, 91, 101
Symmetry numbers, 87
Synchronous processes, 90

Tantalum, diffusion of hydrogen in, 104
Tetramethylguanidine, 97, 139, 179
Tetraphenylporphine tautomerism, 156, 175
Trajectory calculations, 116, 122
Transition state symmetry, 70, 92, 102
Transition state theory, 59
   critique of, 166
Transition state vibrations, 93
Transmission coefficient (classical), 59, 86, 167
Triangular barrier, 29, 39

2, 4, 6-Tri-(1'-adamantyl)phenyl radical, 111
2,4,6-Tri-$t$-butylphenyl radical, 110
2,4,6-Trinitrobenzyl anion, 138
Truncated parabolic barriers, 26
Tunnel correction,
   definition of, 52, 123
   numerical calculation of, 63
Tunnelling frequency, 48
Two-stage processes, 129

Uncertainty principle, 4, 50, 62, 84

Wave-particle duality, 3
Weak quantization, 43
Weber functions, 23

$m$-Xylene, 135

Zero-point energy, 57, 78